Effects of Load and Thermal Histories

On Mechanical Behavior of Materials

Proceedings of a symposium sponsored by the Mechanical
Metallurgy and the Phase Transformation Committees of
TMS-AIME held at the 1987 TMS-AIME Annual Meeting in
Denver, Colorado, February 22-26, 1987.

Edited by

P. K. Liaw

Westinghouse R&D Center
Pittsburgh, Pennsylvania

and

T. Nicholas

AFWAL/MLLN
Wright Patterson AFB, Ohio

A Publication of The Metallurgical Society, Inc.

A Publication of The Metallurgical Society, Inc.
420 Commonwealth Drive
Warrendale, Pennsylvania 15086
(412) 776-9000

Printed in the United States of America.
Library of Congress Catalogue Number 87-42887
ISBN NUMBER 0-87339-028-8

© 1987

Preface

This book is a collection of papers presented at a symposium on "Effects of Load and Thermal Histories on Mechanical Behavior of Materials" sponsored by the Mechanical Metallurgy and the Phase Transformation Committees of The Metallurgical Society of AIME. The symposium was held in Denver, Colorado, on February 25-26, 1987, in conjunction with the Annual Meeting of TMS-AIME. The topics covered a broad area from fatigue crack growth and initiation to phase transformation and fracture over a range of temperatures, load histories, and material and processing variables. The symposium attracted a large audience from all over the United States as well as several foreign countries. Vigorous discussions and technical interchanges among the participants highlighted this symposium.

The papers were divided into three sections: (I) Fatigue Crack Propagation, (II) Isothermal and Thermal-Mechanical Fatigue and (III) Microstructure, Fracture and Damage. A wide range of research topics were covered including both basic research and engineering applications. In Section I, fatigue crack growth behavior was addressed under conditions of variation in load ratio, temperature, test frequency, environment and load spectrum. The role of crack closure was considered in several papers. Experimental aspects and analytical modeling were both addressed. In Section II, fatigue crack initiation was considered under both isothermal and non-isothermal conditions. Several papers addressed critical problems in electronics industry in fatigue of solder materials. Crack initiation in metals was also addressed at both micro- and macro- scales. In Section III, the effect of prior deformation on fracture behavior of steels was considered in several papers. The basic mechanisms which influenced deformation-history dependent properties were investigated and modeled.

The editors would like to thank all of the participants for the success of the symposium and the authors for their contributions to these proceedings. We are grateful to J.M. Wells, R.C. Bates, S. Wood and M.A. Burk for their encouragement during the preparation of the symposium.

Peter K. Liaw
Materials Analysis Department
Westinghouse R&D Center
Pittsburgh, PA 15235

Ted Nicholas
Air Force Wright Aeronautical
Laboratories, AFWAL/MLLN
Wright-Patterson AFB, OH 45433

Table of Contents

Fatigue Crack Propagation

ON TRANSIENTS IN FATIGUE CRACK GROWTH

A. J. McEvily and Y. Zang

Metallurgy Dept. and Institute of Materials Science
University of Connecticut
Storrs, CT 06268

Abstract

The nature of the transients in the rate of fatigue crack growth
which occur as the result of a change in R or ΔK, or as the result of
single or multiple overloads is considered for the ferritic 9Cr-2Mo-0.1C
alloy. In each case crack closure is found to play a dominant role. In
the case of the single overload the plane stress regions at the surface of
a specimen are importantly involved in the retardation mechanism, whereas
the retardation mechanism associated with multiple overloads, a decrease
in R, or a decrease in ΔK is through-thickness in nature. In steady state
crack growth in the absence of closure the crack opening load and the
minimum load in the cycle are found to correspond. A transient
acceleration in the growth rate can occur on increase in R or ΔK, since
the initial crack opening level is below the new minimum load level.

Introduction

Transients in the rate of fatigue crack growth can arise from either a change in loading or environmental conditions, but we will only be concerned with those transients associated with a change in loading in this paper. The most widely investigated transient has been the retardation effect associated with single or multiple overloads, but other transients such as that associated with a change in mean stress at the same ΔK have also been investigated (1). In recent years crack closure has figured importantly in the interpretation of the cause of these transients. For example, Elber (2) and Adams (3) have suggested that the retardation or arrest of a fatigue crack brought on by a reduction of mean load at constant ΔK is due to crack closure. However, Lindley and Richards (1) were unable to detect this enhanced closure under predominantly plane strain conditions by a potential drop technique which had a sensitivity of better than 0.04 mm. Their results for a steel of 415 MPa yield strength indicated that closure occurred above Kmin only at the surface of the specimens in the region of plane stress. These authors also pointed out that the difference in plastic zone shape at the surface and in the interior of a specimen would be expected to influence the nature of the transient. This view was supported by their finding that after an overload the affected region was much smaller in the interior as compared to the surface of a specimen although the minimum rate of growth after an overload was similar in both regions. Since closure in plane strain was not detected over the distance in which the crack growth rate was retarded they proposed that residual stresses were importantly involved in the retardation mechanism in this region.

The duration of the transient before steady state conditions are reestablished depends upon the strength and fracture toughness of the alloy. The strength level is important in establishing the size of the plastic zone, and the toughness will determine whether or not a pop-in event will occur on overloading (4). In a study of a low strength aluminum alloy (yield strength 140 MPa) a pronounced retardation associated with the surface regions after an overload was noted by Matsuoka and Tanaka (5). No pop-in occurred, but because of the high closure developed in the plane stress regions a pronounced tunneling of the crack took place after the overload in the affected region. On the other hand they found that for a steel of medium yield strength (800 MPa) but high toughness, retardation was much less pronounced because of the smaller plane stress plastic zones. It is of interest that on removal of the surface layers by machining that the retardation effect in the case of the aluminum alloy was greatly reduced. However in the case of the steel the crack grew even more slowly after surface removal, an effect which may be attributable to an enhancement of closure in this case due to the removal of surface plane stress regions which served to prop open the crack in the interior. When pop-in and/or tunneling occur the trailing surface layers are in a state of high residual compressive stress on unloading. It has been suggested that as the crack advances through these regions these residual stresses are relaxed and that this relaxation results in an expansion of material in the wake of the crack which increases the crack closure level (6). In this view the maximum retardation should occur in the middle of the affected zone where the extent of relaxation should be a maximum.

Thickness is also a factor in affecting the extent of these transients. As is generally observed, Lindley and Richards (1) found that thickness had no effect on baseline growth rate data for steels tested in the range of thicknesses 1.6 to 20 mm. However, Matsuoka and Tanaka (5) did note a decrease in growth rate with decrease in thickness over a similar range of thicknesses for both the steel and the aluminum alloy mentioned above. The absence of a thickness effect would suggest that the enhanced closure

4

which occurs at the specimen surfaces does not significantly affect the rate of crack growth under steady-state conditions. These surface regions become more important in the case of an overload and both Lindley and Richards, as well as Matsuoka and Tanaka found that as the specimen thickness was reduced the extent of retardation for a given overload was increased.

With this background we are now ready to focus attention on several aspects of the broad topic of transient effects due to changes in load cycling. In particular we will consider the effect of a change in R at constant ΔK, the effect of a change in ΔK at constant R, and the comparison of the nature of closure for single and multiple overload events.

Transients Due to a Change in R

The material selected for the test program to be described was a 9Cr-2Mo ferritic steel (yield strength 500 MPa). Ferritic steels exhibit some interesting fatigue crack growth characteristics. For example, as shown in Fig. 1, the rate of fatigue crack growth at 538°C in vacuum is independent of mean stress, an effect due to the absence of crack closure (7). It is also noted that the rate of fatigue crack growth is the same as for other ferritic steels although the alloys differ with respect to initial mechanical properties. Further, over a range of frequencies down to 0.3 Hz the rate of crack growth was found to be independent of frequency, an indication that over this frequency range creep per se did not play a significant role in these tests. Since the rate of growth was independent of R the opportunity presented itself to interpret the effect of a change in R on the subsequent growth rate in a straightforward manner. Results

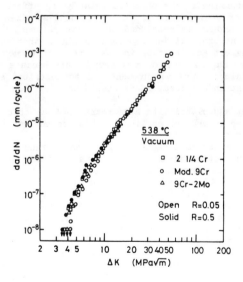

Fig. 1   Fatigue crack growth rate (da/dN) vs. range of stress intensity factor (ΔK) for 9Cr-2Mo as well as mod. 9Cr-1Mo and 2.25Cr-1Mo ferritic steels in vacuum at 538°C (7).

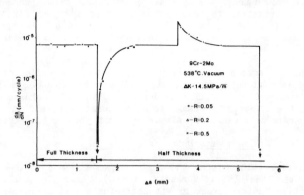

Fig. 2   The effect of a change in R level on the rate of fatigue crack
         growth in 9Cr-2Mo steel at 538°C in vacuum. Prior to changing
         to an R of 0.05 from an R of 0.5, 1/4 of the thickness was
         machined from each surface to eliminate any surface-related
         effects (8).

of a typical test sequence are shown in Fig. 2 (8). In this sequence ΔK
was held constant and only the value of R was changed. Before propagating
the crack at the lower R ratio the 6.25 mm thick compact specimen was
removed from the test apparatus and 1.5 mm were machined from each specimen surface to determine if the observed retardation effect was a through
thickness effect or not. Tests of full thickness specimens resulted in
similar behavior, and it is concluded that this retardation effect is
full-thickness in nature. It is also seen in Fig. 2 that an increase in R
results in a transient acceleration in the rate of fatigue crack growth.
Similar tests were carried out at room temperature because it was simpler
to carry out certain closure measurements to be described although some
closure was present under steady state conditions.

A main effect associated with a decrease in R at constant ΔK is an
associated decrease in mean stress level. This decrease is responsible for
the observed retardation on decrease of R, for it leads to the development
of closure due to an overall reduction in elastic tensile strain and a
consequent clamping action on the crack tip as the mean stress is reduced
(8). However the clamping action is exerted over only a very short
distance, i.e., some 100 microns or less behind the crack tip at zero
load, and because of the short distance involved this closure was not
detected with a front-face clip gauge mounted on the compact specimen.
Strain gauges mounted close to the crack tip did indicate that closure
occurred after the transition from a high to a low R level, but even this
technique was not considered to be sensitive enough. In order to obtain
the requisite sensitivity a replication procedure was adopted. In this
procedure the specimen was subjected to a series of increasing loads for
several crack lengths within the transient region and a replica of the
crack tip region was taken at each load level using acetate film moistened
with acetone. These replicas were coated with a fine layer of gold and
then examined in a scanning electron microscope. The specimen surface was
removed by polishing in order to study the variation of the closure level
with depth into the specimen. Closure levels at the surface were some 20%

6

higher than below the surface with the transition occurring at a depth given approximately by

$$\frac{K^2}{2\pi\sigma_y^2}$$

which relates to the plane stress plastic zone size. The closure level was constant at greater depths into the specimen. Fig. 3 shows examples of appearance of the replicas as viewed in the SEM. It is noted that at the lowest load level shown the crack is open some 50 microns behind the actual crack tip. As the load level increases the crack opening point moves toward the tip. It is significant that the initial tip opening level corresponds to the minimum load level of the higher R level.

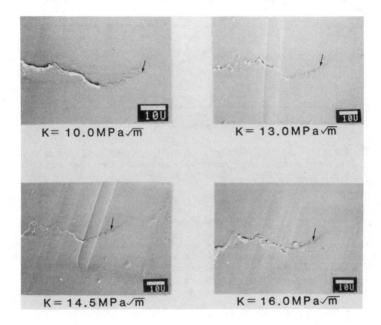

Fig. 3   SEM photomicrographs of fatigue crack tip region in interior of specimen during crack opening process. Prior cycling was at R = 0.5 with Kmin = 14.5 MPa√m. As K increases the opening point moves toward actual tip which is marked by arrows. Crack tip opening occurs at Kop = Kmin.

The results of such crack tip opening levels as the crack grows through the transient zone are shown in Fig. 4. These results pertain to the closure level beneath the surface plastic zone. As the closure level changes so also does the rate of crack growth. The crack growth rates through the transient region correspond to those expected on the basis of ΔKeff. When the crack tip opening level decreases to the minimum load of the reduced R level the rate of crack growth regains its R-independent value and the transient period is at an end. We note that if the R level is then returned to its initial value there is initially an acceleration in growth rate before the rate again settles into its steady-state value.

7

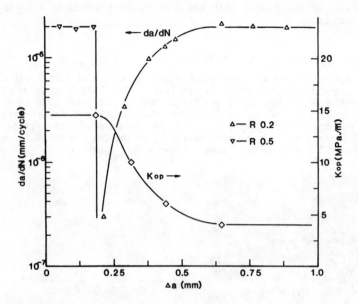

Fig. 4   Variation of the opening load in the transient region and the
          corresponding crack growth rate after a change in R from 0.5 to
          0.2.

ΔK Transients

Transients similar in nature are observed on change in ΔK at constant
R as shown in Fig. 5. In passing it is noted that in obtaining crack
growth rate data as a function of ΔK in a decreasing ΔK test it is of
course important that data be obtained only after the transient period has
terminated.

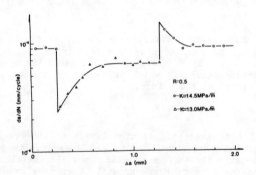

Fig. 5   Effect of a change in ΔK level at an R of 0.5 on the rate of
          fatigue crack growth in 9Cr-2Mo steel at 538°C in vacuum (8).

Effect of Single and Multiple Overloads

The effect of either single or multiple overloads on the rate of fatigue crack growth in the 9Cr-2Mo steel is indicated in Fig. 6. In these tests the minimum stress intensity factor was kept at 0.5 MPa√m. The upper maximum level (overload level) was 15.8 MPa√m and the lower maximum level was 10.5 MPa√m, corresponding to a 50% overload. (For multiple overloads with the same minimum level but with the maximum changed from 21 to 10.5 MPa√m crack arrest occurred, i.e., no growth in 10^6 cycles was observed). Two stages of testing were involved in order to compare the crack growth behavior in the interior of the specimen with that at the surface. After the overload crack growth rate conditions had been determined for the full thickness, 6.25 mm specimen, the specimen was again tested to reestablish the initial steady state conditions. Then after the single or multiple overloading, 1.5 mm were machined from each specimen surface and the test continued at the lower range. In the tests which involved the machining away of the surface layers it is assumed that there is no effect on the closure level of the remaining portion of the specimen as a result of surface removal, but as the results of Matsuoka and Tanaka (5) indicate this may not always be the case. With this caveat, the results obtained for the various test conditions are summarized in Fig. 6.

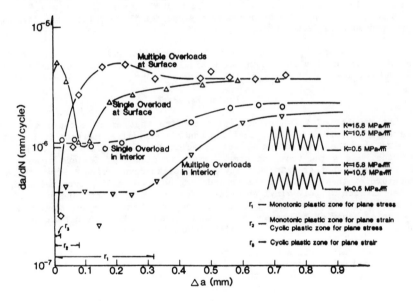

Fig. 6   Comparison of the rate of fatigue crack growth at the surface and in the interior of 9Cr-2Mo specimen after 50% multiple and single overloads under ambient conditions.

After the multiple load sequence the retardation effect on the full thickness specimen is immediate upon reduction in maximum load, and the rate returns to its normal value in a distance of 400 microns. When machining is done after the multiple overloads the steady state rate is reestablished in 600 microns. There is some scatter for this case which may relate to near-threshold crack propagation characteristics, but the curve through the data points is drawn to indicate immediate retardation after the multiple overloads. It is also noted that because of crack front curvature the crack in the interior is initially one millimeter more advanced than it is at the surface. After machining it is seen that the rate of crack growth is slightly less in the steady state than before machining. This difference appears to be related to a higher closure level after machining (3.5-4 MPa√m) than before (3 MPa√m).

After the single overload the crack at the surface of the full thickness specimen exhibited a delay in retardation before regaining the steady state value. The minimum crack growth rate occurred after a 100 micron crack advance. The results obtained after machining following the overload indicate an immediate retardation. For both surface and interior regions the crack growth rate returned to the steady state value in about 500 microns, with the steady state crack growth rate for the thinner specimen again being less than for the full thickness specimen.

For purposes of reference the monotonic plastic zone size is taken to be $K^2/\pi\sigma_y^2$, and for a yield stress of 500 MPa and a Kmax of 15.8 MPa√m the calculated plane stress plastic zone size size would be 300 microns. If the cyclic yield stress is taken to be 1000 MPa then the zone would be 1/4 as large. For plane strain conditions each of these values is estimated to be reduced by a factor of four.

It is also noted that the extent of retardation is greater in the case of multiple overloads than for a single overload. This is true for both the surface and interior of the specimens.

Concluding Remarks

The results of this investigation confirm that crack closure is importantly involved in the transients associated with a decrease in R or ΔK as well as the retardation effect associated with single or multiple overloads. In going from one steady state condition of fatigue crack propagation to another transient involves a shake-down period of transition. The nature of these transitional periods can be more readily examined with a material such as poly-vinyl chloride (PVC) than with a metal because of the much larger deformations that can be imposed. For example, the crack tip opening behavior before and after an increase in Kmin with Kmax held constant is shown in Fig. 7. Immediately after the change in loading the crack tip opening characteristics are the same as for the previous loading, but these characteristics quickly change to reflect the new cyclic conditions. It is noted that the new crack opening level now corresponds to the minimum load in the cycle and that the opening at Kmax is greatly reduced and corresponds to the particular range of ΔK applied independent of the mean stress. Such experiments indicate why the rate of crack growth shown in Fig. 1 is independent of R.

Similar experiments have shown that immediately upon increase in R at constant ΔK the extent of plastic deformation at the crack tip is relatively large but decreases rapidly with further cycling. Such behavior is responsible for the acceleration in the rate of fatigue crack growth noted on increase in R or ΔK, Figs. 2 and 5. In these cases the initial crack opening level is below the minimum level for the new cyclic conditions and this leads to a larger crack tip opening displacement for a given ΔK than if the opening level were at the minimum level of the cycle.

10

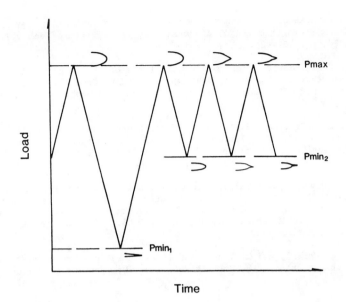

Fig. 7   Variation in crack opening characteristic with change in loading
conditions based upon observations of PVC specimens. Note that
in steady state Kmin and Kop are equal.

The results also indicate that the effects of a single overload are
more pronounced at the surface of the specimen than in the interior, an
indication that the plane stress plastic zones are importantly involved in
the retardation mechanism for this type of overload. In the case of the
multiple overloads the retardation mechanism is through-thickness in
nature and is associated with a clamping action developed at the crack tip
upon reduction in mean stress level.

Acknowledgement

The support received for this study from the U. S. Department of
Energy, Office of Basic Energy Sciences, Division of Materials Sciences,
GRANT DE-FG02-84ER45109 is gratefully acknowledged.

References

1.   T. C. Lindley and C. E. Richards, The Relevance of Crack Closure to
Fatigue Crack Propagation, Mats. Science and Eng., vol 14, 1974, pp
281-293.

2.   W. Elber, Fatigue Crack Closure under Cyclic Tension, Eng. Fracture
Mech., vol 2, 1970, pp 37-45.

3.   N. J. I. Adams, Fatigue Crack Growth at Positive Stresses, Eng.
Fracture Mech., vol 4, 1972, pp 543-554.

4. C. Bathias and M. Vancon, Mechanisms of Overload Effect on Fatigue Crack Propagation in Aluminum Alloys, Eng. Fracture Mech., vol 10, 1978, pp. 409-424.

5. S. Matsuoka and K. Tanaka, The Influence of Sheet Thickness on the Delayed Retardation Phenomena in Fatigue Crack Growth in HT80 Steel and A5083 Aluminum Alloy, Eng. Fracture Mech., vol 13, 1980, pp 293-306.

6. A. J. McEvily and K. Minakawa, Crack Closure and Fatigue Crack Growth Under Variable Amplitude Loading, in Fundamental Questions and Critical Experiments on Fatigue, ASTM STP , ed. by J. T. Fong and R. P. Wei, to be published.

7. H. Nakamura, K. Murali, K. Minakawa, and A. J. McEvily, Fatigue Crack Growth in Ferritic Steels as Influenced by Elevated Temperature and Microstructure, in Proc. Int. Conf. on Microstructure and Mechanical Behavior of Materials, Xi'an, China, 1985. To be published by EMAS, Warfield, England.

8. A. J. McEvily and Z. Yang, Transients in Fatigue Crack Growth due to a Change in ΔK or R, Scripta Met., vol 20, 1986, pp 1781-1784.

ELEVATED TEMPERATURE FATIGUE CRACK PROPAGATION

AFTER SUSTAINED LOADING

K.-M. CHANG

Materials Laboratory
General Electric Company
Corporate Research and Development
Schenectady, New York 12301

Abstract

Fatigue crack propagation (FCP) in superalloys at intermediate tempera-tures is well known to show a strong frequency dependence, and the oxidation environment plays a major role on the crack growth rate. An experimental technique has been developed to understand and to evaluate the environmental attack at the crack tip under the loading stress. Inconel 718 alloy was used as the model alloy. After a given period of sustained loading, the crack was fatigued by a normal fatigue frequency under a constant cyclic stress inten-sity. Crack growth rate was measured every cycle until the crack growth rate reached to the normal value. The "damaged" zone induced by sustained loading resulted in a much faster crack growth rate compared to the normal value. The size of "damaged" zone depended upon the sustained load, the loading period, and the environment for a given temperature.

13

Introduction

Fatigue crack propagation (FCP) in the high-strength superalloys at elevated temperatures has received extensive attention in the past decade. Aircraft engine and gas turbine industries have already accepted the fracture mechanism methodology to design and to manage the life of critical components (1,2). As a result, many FCP data were determined under various testing procedures and conditions (3-8).

One of the most interesting observations in elevated temperature FCP is the time-dependence of the crack growth rate (9). Under a given cyclic stress intensity, ΔK, the crack growth rate, da/dN, in superalloys maintains a constant value at lower temperatures, no matter what different cyclic frequencies or waveforms of fatigue cycles are used. This testing regime is called the cycle-dependent FCP. However, when the testing temperature is increased, da/dN is no longer a simple function of ΔK only. Both cyclic frequency and waveform can affect significantly the FCP behavior. Time-dependent FCP can be accelerated substantially by slowing down the fatigue frequency, or by imposing a hold-time at the maximum load of fatigue cycles.

Dynamic embrittlement of air environment at elevated temperatures has been proposed to explain time-dependent FCP, since time-dependence of da/dN becomes negligible when a superalloy is tested in a high vacuum chamber (8-10). Air environment is believed to interact with the crack tip under the fatigue loading cycle, and a time period is required to allow the crack tip embrittled. The loading stress is also a necessary element to cause such a dynamic embrittlement.

This work intends to develop an experimental technique for the exclusive investigation of the dynamic embrittlement associated with time-dependent FCP. The "damaged" zone in the crack tip, which is induced by a sustained loading, can be characterized by a mechanical method, i.e., the crack advances every fatigue cycle after the sustained loading. The size of "damaged" zone depended upon the loading period, the loading stress intensity, and the temperature.

Experimental

Materials

Table 1 lists the chemical composition of Inconel 718 alloy that was employed as the model alloy for this study. All samples were obtained from a commercial premium grade forging that is used for aircraft engine hardwares. A standard heat treatment: 1025°C/1hr + 720°C/8hr + 620°C/10hr, was applied to all sample blanks before the specimens were machined.

Table 1.   Chemical composition of 718 alloy.

Ni	Cr	Fe	Mo	Nb+Ta	Ti	Al	C	B
bal.	19.0	18.0	3.0	5.1	0.9	0.5	0.06	0.006

Fatigue Specimen

Fatigue crack propagation tests were performed by employing the single-edge-notched (SEN) sheet specimen, Figure 1. A reversible dc current was applied by a power supplier through both heads of the specimen. The crack length was monitored by a pair of potential probes mounted on the front edge of the specimen gage across the pre-machined notch. The measured dc potential drop at any crack length was normalized and converted into the corresponding crack length by a single analytical relation, namely Johnson's equation (11):

14

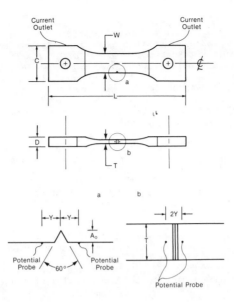

Figure 1 - Schematic drawing of the single-edge-notched (SEN) specimen for fatigue crack propagation (FCP) test.

$$A = \frac{2W}{\pi} \times \cos^{-1}\left[\frac{\cosh(\pi Y/2W)}{\cosh[(U/U_o)\cosh^{-1}[\cosh(\pi Y/2W)/\cos(\pi A_o/2W)]]}\right] \qquad (1)$$

where A and Ao denoted the actual and initial crack lengths, respectively; U and Uo were the updated and initial measured potential drops, respectively; W and Y represented, respectively, the specimen width and one half of the potential probe span as shown in Figure 1. Subsequently, the stress intensity factor was calculated by Tada empirical equation (12):

$$K = \frac{P}{WT} \times \sqrt{\pi A} \times f(\frac{A}{W}) \qquad (2)$$

$$f(\frac{A}{W}) = \sqrt{\frac{2W}{\pi A} \times \tan(\frac{\pi A}{2W})} \times g(\frac{A}{W})/\cos(\frac{\pi A}{2W}) \qquad (3)$$

$$g(\frac{A}{W}) = 0.752 + 2.02(\frac{A}{W}) + 0.37[1-\sin(\frac{\pi A}{2W})] \qquad (4)$$

where P denoted the applied load, and T represented the specimen thickness. The accuracy of the above equations has been confirmed better than 0.5% providing that no bending movement is applied.

Testing Procedures

The SEN specimen was tested in a closed-loop, servo-hydraulic test machine equipped with a resistance furnace. After heating to the testing temperature (400 to 650°C), the specimen was precracked with a low stress intensity ΔK and 10 Hz sinusoidal cycles up to at least the notch depth. A microprocessor continuously recorded the testing data, and in real time calculated the crack length, the stress intensity factor, and the linear regression analysis of crack growth rate. The servo-hydraulic test machine was modified to be controllable by the microprocessor, and the testing mode could be automatically changed when the crack reached the preset value.

The tests were carried out by using the constant stress intensity mode for a given temperature. A "normal" fatigue frequency of 3.0 cpm (20 seconds per cycle) was employed so that the crack length can be determined every cycle. The minimum-to-maximum load ratio is selected at R = 0.05. When the crack developed a stable growth rate, da/dN, a sustained load of a specific stress intensity and of a specific time period was applied. During the sustained loading, the crack length was constantly checked to monitor the crack growth until the end of sustained loading period. As soon as the sustained loading was terminated, the normal fatigue cycles was reapplied, and the crack advance was recorded cycle by cycle.

The specimen was broken into halves after the testing was finished. The measurement of beach marks appearing on the fracture surface was made for the crack length calibration. The prediction of Johnson's equation agreed well with the actual crack length within 10% error. The ASTM seven-point incremental polynomial method was employed to calculate crack growth rate (13).

Materials Characterization

Metallographic samples were prepared using conventional mechanical grinding and polishing procedures according to the standard laboratory process. An etching solution consisting of 10 ml HCl, 10 ml HNO_3, and 30 ml H_2O_2 was used to reveal microstructure. Fractography was examined under a scanning electron microscope for each sample tested under different testing conditions.

Results

Time-Dependent Behavior:

Figure 2 shows the crack growth rate of standard heat-treated Inconel 718

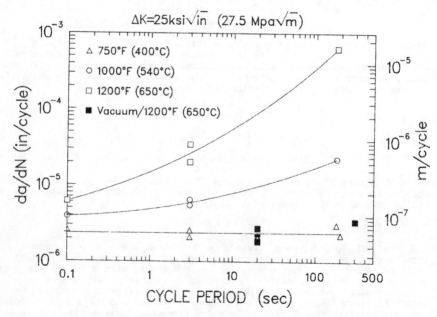

Figure 2 - The relationship between crack growth rate and period of fatigue cycles in the temperature range of 400 to 650°C for Inconel 718.

as a function of fatigue frequency at different temperatures under a fixed cyclic stress intensity, $\Delta K = 27.5$ MPa√m̄. Two regimes of FCP behavior are clearly observed. At 400°C, the alloy shows a constant da/dN under a given ΔK. The crack growth is completely determined by the cyclic stress intensity, independent of fatigue frequency. In this cycle-dependent regime, crack growth rate is believed to be insensitive to the variation of microstructure and alloy chemistry. As the temperature increases, the crack growth starts to accelerate substantially even if ΔK is kept at the same value. At 540°C, da/dN is no longer the same for different fatigue frequency, and increases with the cycle period. Such a time-dependent behavior is aggravated by increasing temperature; two order of magnitude increment in the crack growth rate is detected by simply slowing down the fatigue frequency at 650°C.

Crack Growth After Sustained Loading:

In the time-dependent regime, a faster fatigue crack growth rate was observed after sustain loading. One example is given in Figure 3. The test-ing temperature is 590°C, and the cyclic stress intensity is 26.5 MPa√m̄. After loading at 28 MPa√m̄ for 1000 seconds, the crack starts to grow at a rate 10 times faster than the normal da/dN (before sustained loading). The stress intensity of sustained loading, K_{sl}, is set at the same value as the maximum stress intensity of fatigue cycles, K_{max}, so that no incubation is induced by the plastic zone size in front of crack tip. The crack growth rate monotoni-cally decreases as the crack propagates forward, and eventually reaches the normal value. The distance to stabilize da/dN in this case is about 0.090 mm.

Loading Period Effect. Figure 4 shows the crack growth curves after sustained loading for various periods (300 to 10000 seconds) at 590°C. The cyclic stress intensity is $\Delta K = 26.5$ MPa√m̄ ; the sustained loading stress intensity is equal to the maximum stress intensity, K_{max}, of the fatigue cycles. Two apparent features about the crack growth after sustained loading are observed in Figure 4:

a) The initial crack growth rate after sustained loading, which corresponds

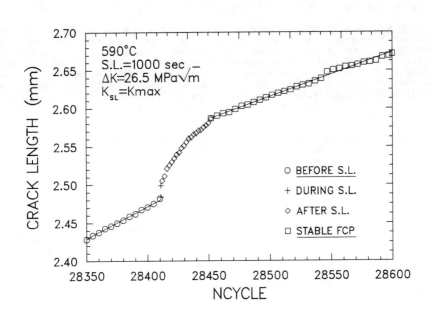

Figure 3 - Crack growth behavior with a sustained loading.

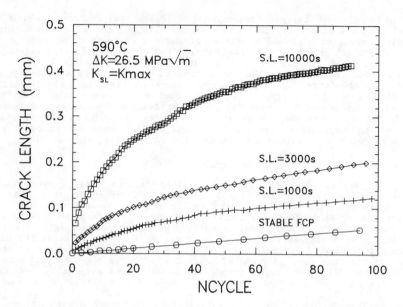

Figure 4 - Crack growth curves after sustained loading for different loading periods.

to the slope of the crack growth curve at the starting point, increases remarkably with the loading period.
b) The distance to reach the normal da/dN, i.e., stable crack growth rate, also increases with the loading period.

Temperature Effect.   Increasing testing temperature always accelerates the crack growth as seen in Figure 5.   The temperature effect becomes much more pronounced in the time-dependent regime.   Therefore, the normal da/dN values are different even though ΔK is maintained the same.   Figure 5 shows crack growth behavior after sustained loading at 26.5 MPa$\sqrt{m}$ for 1000 seconds at three temperatures, 540°, 590°, and 650°C.   Increasing temperature has the similar effect qualitatively as increasing loading period; i.e., a high initial crack growth rate and a long stabilizing distance (instead of number of fatigue cycles) are observed.

Loading Stress Effect.   Three K_{sl} levels, 24, 28, 39 MPa$\sqrt{m}$ have been employed to study the loading stress effect on the subsequent crack propagation.   The loading period is chosen to be 3000 seconds, and the testing temperature is 590°C.   To avoid the incubation, the fatigue cycles after sustained loading is set such that $K_{sl} = K_{max}$ with a fixed R = 0.05.   The normal da/dNs are then different for different ΔK.

The resulting crack growth curves are shown in Figure 6.   As expected, the higher the loading stress intensity is applied, the faster the crack grows initially, and the longer the distance is required to stabilize da/dN.   However, the influence of loading stress does not seem to be as significant as that of loading period and temperature.   This observation is in agreement with many prior data (10,14).   When the crack growth rate, da/dN, are plotted as a function of the cyclic stress intensity, ΔK, the stage II crack growth rate can be filled by Paris' power law,

$$\frac{da}{dN} = C \times (\Delta K)^n \tag{5}$$

18

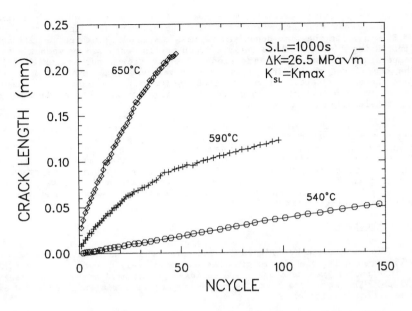

Figure 5 - Crack growth curves after sustained loading at different testing temperatures.

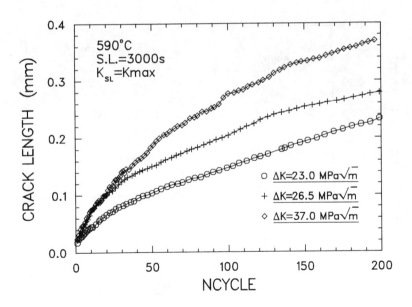

Figure 6 - Crack growth curves after sustained loading with different loading stress intensities.

In log-log plotting, crack growth curves are usually represented as a straight line by linear regression analysis, and the slope of the line is the exponent of Eq. (5). In the time-dependent regime, the crack growth curve will move up (increasing da/dN) when the fatigue frequency is decreased or when the testing temperature is increased. The slope of crack growth curves, which reflects the dependence of loading stress, has never undergone any dramatic change. Nevertheless, a certain level of loading stress is necessary to affect the crack growth behavior. A long period holding with zero stress does not cause the acceleration of crack growth.

Crack Growth During Sustained Loading. The crack also propagates during the sustained loading; the situation is very similar to so-called "creep" crack growth. Figures 7 and 8 shows the results obtained for different temperatures and for different loading stress intensities, respectively. Except for a short period at the beginning of sustained loading, the crack grows linearly as a function of time, and the data can be fitted very well by a straight line. Again the temperature results in a more substantial effect than the loading stress intensity.

Microstructure

The metallography, Figure 9, indicates that the solution annealing at 1025°C can eliminate the forging history completely. A fully recrystallized grain structure was developed, and equiaxed grains have a grain size of ASTM 7 (35 μm). The structure seems to be isotropic; no stringer is observed except some isolated carbides.

Different fracture modes (Figure 10) are found on the fracture surface of broken specimens. Conventional fatigue facets with continues striations (Figure 10(a)) are developed in the precracking, which employes a fast fatigue frequency, 10 Hz. Depending upon the testing temperature, the "normal" frequency, 3 cpm, introduces the intergranular failure that is the typical

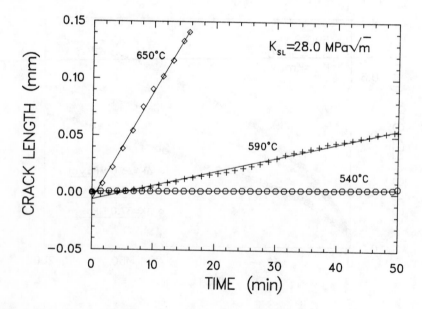

Figure 7 - Crack growth during the sustained loading at different testing temperatures.

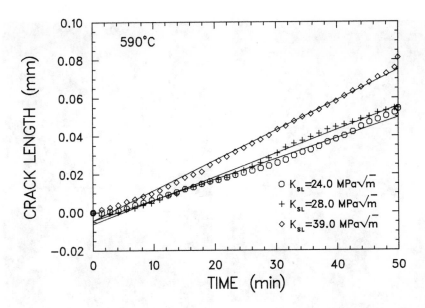

Figure 8 - Crack growth during the sustained loading with different loading stress intensity.

failure mode for time-dependent FCP. At 540°C, a mixture of transgranular and intergranular fracture modes is shown (Figure 10(b)). As the temperature is raised to 590°C, only a small portion of fracture surface exhibits transgranular facets (Figure 10(c)). Complete intergranular failure is detected at 650°C (Figure 10(d)).

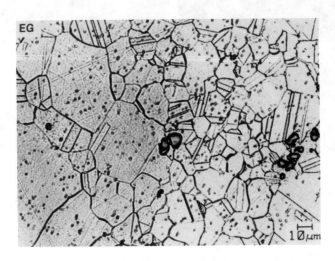

Figure 9 - Grain structure of the standard heat-treated Inconel 718.

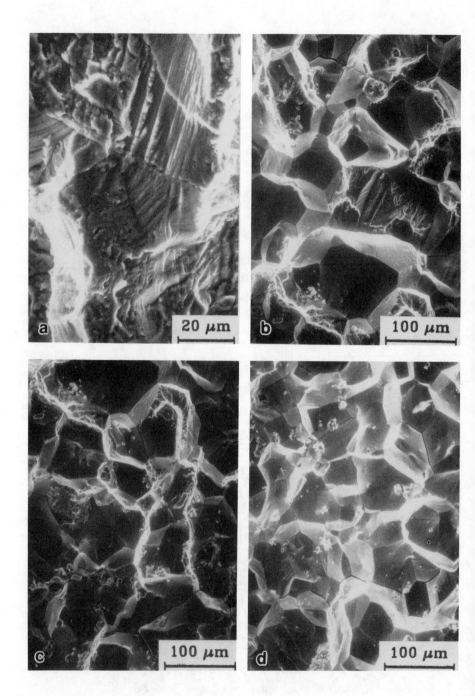

Figure 10 - Fatigue fracture modes observed at: (a) 540°C, 10 Hz;
(b) 540°C, 3 cpm; (c) 590°C, 3 cpm; (b) 650°C, 3 cpm.

Discussion

Damaged Zone of Dynamic Embrittlement

The above results provide a general foundation for better understanding of high-temperature cracking problem in high-strength superalloys. Theoretically, linear elastic fracture mechanism (LEFM) should be able to characterize the cracking behavior of a material satisfactorily. This is true in the cycle-dependent region; crack growth rate is simply determined by the cyclic stress intensity. In other words, da/dN is one of the unique material properties, just as alloy strength is.

To consider the time-dependent FCP, the testing environment plys a major role in the measured da/dN. Included in Figure 2 are some vacuum data of Inconnel 718 tested at 650°C. These data are in consistence with the cycle-dependent values, which are measured at 450°C in air. In contrast to 650°C air data that show a strong dependence of fatigue frequency, the vacuum data maintain at a constant value (cycle dependent only). Therefore we may conclude that the air environment, persumed oxygen, interacting dynamically with the material in front of the crack tip causes the time-dependence of fatigue crack propagation. Such environmental embrittlement occurs through the crack tip, and is localized in a zone just ahead of the crack tip. The "damaged" zone no long represents the original bulk material and allows the crack to grow at an accelerated rate under the subsequent fatigue cycles.

Figure 3 demonstrates the existence of the damaged zone clearly. Fatigue crack grows at a specific rate under a constant cyclic stress intensity; a sustained loading is applied to generate a damaged zone through dynamic embrittlement by air environment. A fast crack growth rate is observed when the fatigue cycle is reassumed. As the crack is growing, da/dN slows down until the normal value is reached, and then the crack grows at the original rate again. The damaged zone is obviously the distance from the point that the sustained loading is removed to the point that da/dN returns to normal.

The damaged zone can be easily determined by plotting the crack growth rate as a function of crack length after sustained loading. One example is given in Figure 11. In the plot, da/dN has a logarithmic axis, and the zero of linear crack length axis is defined at the crack length just after sustained loading. The data points can be divided into two parts; and each part can be fitted by a straight line. In the first part, crack growth rate monotonically decreased in a linear fashion. In the second part, a horizontal line suggests the constant crack growth rate, which is corresponding to the "normal" da/dN value. The intersection of two lines in Figure 11 determines the size of the damaged zone induced by the sustained loading.

Mathematically, the first part of Figure 11 can be formulated as

$$\log(\frac{da}{dN}) - \log(\frac{da}{dN})_o = C \times (A\text{-}D) \tag{6}$$

where $(da/dN)_o$ is the normal value of the second part, A is the crack length after sustained loading, and two constants are: D representing the size of the damaged zone, and C being the decade constant, which has a negative value. Eq. (6) can be rearranged,

$$\frac{(da/dN)}{(da/dN)_o} = \exp(C \times (A\text{-}D)) \tag{7}$$

Now Eq. (7) implies some interesting physical meaning of the damaged zone. On the left-hand side of Eq. (7), the ratio of crack growth rate to the normal value reflects in the mechanical sense the degree of embrittlement (the damage) at a specific point in front of the crack tip. On the right-hand side, the damage is exponentially decaded from the crack tip to a distance D (damage zone) with a decade constant C. The microstructural characterization of such an exponentially decaded damage requires further investigation.

23

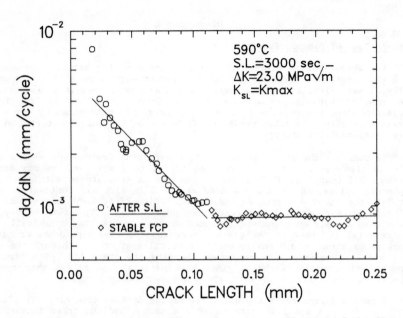

Figure 11 - Measured fatigue crack growth rates as a fuction of the distance after sustained loading.

Time-Dependent Fatigue Crack Propagation

One of the major difficulties of employing the fracture mechnism methodology to predict component fatigue life is how to define the appropriate simulation of fatigue cycles during service. A straightforward case is the mixture of cyclic and static loading. Many proposals based on the linear superposition of "fatigue" (cyclic) and "creep" (static) crack growth rates have been suggested (15,16). However, these simple-minded models can only be valid when one type of crack growth mechanism, either cyclic or static, is dominant. The basic assumption is that there is no interaction between two types of loading. Figure 3 shows an apparent evidence to disprove these models.

A phenomenological approach to time-dependent FCP behavior must employ the idea of the damaged zone as the foundation. Since the degree of embrittlement in the damaged zone is the one and only one parameter to decide the crack growth rate, a universal index, which can be called as time-dependent factor, should be used for every FCP testing condition. The time-dependent factor will include cycle frequency (time), waveform (stress), and temperature; the detailed formula requires an extensive study on the formation of damage zone. A high value of the time-dependent factor means high temperature, and/or slow frequency, and/or high stress level. Hence, for a given cyclic stress intensity, the crack growth rate becomes a function of the time-dependent factor as shown schematically in Figure 12. When the time-dependent factor is small, crack grows at a constant rate, which is only determined by cyclic stress intensity, ΔK. This regime is called cycle-dependence. As the time-dependent factor increases, crack growth rate is no longer a constant and increases with the time-dependent factor. In this time-dependent regime, two parameters are of most interest. One is the offset value of the time-dependent factor for the beginning of time-dependent FCP. The other is the

24

High Strength Superalloys
Fatigue Crack Propagation

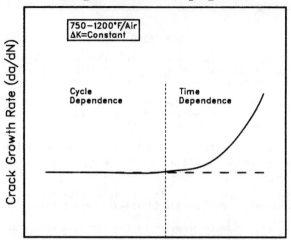

Time Dependent Factor

Figure 12 - Fatigue crack propagation (FCP) behavior of high-strength
superalloys at elevated temperatures.

degree of time-dependence when crack growth becomes time dependent. Both
parameters are materials properties only and can be controlled through the
modification of alloy chemistry and microstructure.

Conclusion

The influences of sustained loading on the crack growth behavior has been
investigated in a high-strength superalloy, Inconnel 718, at the temperatures
up to 650°C. Dynamic embrittlement occurs in front of crack tip through the
interaction with the air environment, and a certain size of the damaged zone
develops to cause subsequent fast crack growth rate. Some important features
and implications of the damaged zone follow:

a) The degree of air embrittlement within the damaged zone, measured by the
   ratio of the accelerated crack growth rate to the normal value, is
   exponentially decaded from the crack tip after sustained loading.
b) The initial crack growth rate in the damaged zone is increased with the
   temperature, the loading period, and also the stress intensity of
   sustained loading.
c) The above factors of sustained loading affect the size of the damaged
   zone as well.
d) A unique time-dependent factor, based on the concept of the damaged zone,
   is expected to characterize the fatigue crack propagation behavior of
   high-strength superalloys at elevated temperatures.

Acknowledgments

The author is gratefully indebted to E.H. Hearn, B.J. Drummond, and H.

Moran for providing direct assistant of experimental work. Many thanks are also due to L.J. Beha, D.C. Lord, and D.A. Catharine for their technical support in FCP tests. Helpful discussions with Dr. M.F. Henry and Dr. M.G. Benz are greatly appreciated.

References

1. G.W. Meetham, "Materials for Advanced Gas Turbines," in High Temperature Alloys for Gas Turbines and Other Applications 1986, ed. W. Betz (Dordrecht, Holland: D. Reidel Publ. Co., 1986), 1-18.

2. T. Nicholas and J.M. Larson, "Damage Tolerance requirements and Implications for Materials Development," (Paper presented at 1986 TMS Fall Meeting, Orlando, FL, 7 October 1986), 47.

3. H.D. Solomon and L.F. Coffin, Jr., "Effects of Frequency and Environment on Fatigue Crack Growth in A286 at 1100°F," Fatigue at Elevated Temperatures, (ASTM STP 520, Amer. Soc. Test. Mat. 1973) 112-122.

4. V. Shahani and H.G. Popp, "Evaluation of Cyclic Behaviour of Aircraft Turbine Disk Alloys," (NASA Report CR-159433, June 1978).

5. B.A. Cowles et al., "Cyclic Behavior of Turbine Disk Alloys at 650°C," J. Eng. Mat. Tech., 102 (Oct. 1980) 356-363.

6. B.A. Cowles, D.L. Sims, and J.R. Warren, "Evaluation of Cyclic Behaviour of Aircraft Turbine Disk Alloys, Part II" (NASA Report CR-165123, July 1980).

7. J. Gayda and R.V. Miner, "The Effect of Microstrcture on 650°C Fatigue Crack Growth in P/M Astroloy," Met. Trans. A, 14A (Nov. 1983) 2301-2308.

8. H.H. Smith and D.J. Michel, "Effect of Environment on Fatigue Crack Propagation Behavior of Alloy 718 at Elevated Temperatures," Met. Trans, A, 17A (Feb. 1986) 370-374.

9. K.-M. Chang, "Improving Crack Growth Resistance of IN 718 Alloy through Thermomechanical Processing," (GE-CRD Report No. 85CRD187, Oct. 1985).

10. K.-M. Chang, "Time-Dependent Fatigue Propagation in Inconel 718 Superalloys," (GE-CRD Report No. 86CRD228, Feb. 1987).

11. K.-H. Swalbe and D. Hellmann, "Application of the Electrical Potential Method to Crack Length Measurements Using Johnson's Formula," J. Eng. Mat. Tech., JTEVA 9 (3)(May 1981) 218-221.

12. H. Tada, P. Paris, and G. Irwin, The Stress Analysis of Cracks Handbook, (Del Research Corp. 1973) 2.10-2.12.

13. "Constant-Load-Amplitude Fatigue Crack Growth Rates Above 10^{-8} m/Cycle," 1980 Annual Book of ASTM Standard, Part 10, E 647-78T.

14. K.-M. Chang, "Grain Size Effects on the Fatigue Crack Propagation of High-Strength Superalloys," (GE-CRD Report No. 86CRD201, Oct. 1986).

15. J. Gayda, T.P. Gabb, and R.V. Miner, "Fatigue Crack Propagation of Nickel-Base Superalloys at 650°C," (NASA Report TM-87150, Sep. 1985).

16. A. Saxsena, "A Model for Predicting the Effect of Frequency on Fatigue Crack Growth Bebaviour at elevated temperature," Fat. Frac. Eng. Mat. Str., 3 (3)(1980) 247-255.

ROLE OF CRACK CLOSURE IN CRACK RETARDATION

IN P/M AND I/M ALUMINUM ALLOYS

J. J. Kleek[a] and T. Nicholas[b]

[a]Air Force Wright Aeronautical Laboratories
Materials Laboratory, AFWAL/MLLS
Wright-Patterson AFB, OH 45433-6533

[b]Air Force Wright Aeronautical Laboratories
Materials Laboratory, AFWAL/MLLN
Wright-Patterson AFB, OH 45433-6533

Abstract

Two aluminum alloys, P/M CW67 and I/M 7475 were subjected to both constant amplitude fatigue and a simple fatigue spectrum consisting of constant amplitude loading with a periodic overload of 80 percent applied every 8000 cycles. Fatigue crack growth rates were derived from crack length data obtained using crack tip opening displacement (COD) compliance measurements on C(T) specimens. Several additional tests were conducted under constant ΔK conditions to observe the growth rates and closure loads immediately after a single overload.

Crack growth retardation due to periodic overloads was observed in both alloys for the entire range of ΔK values investigated. The crack growth rate in P/M CW67 was higher than in I/M 7475 under both constant amplitude and periodic overload loading. The differences were attributed to differences in closure loads which, in turn, were due to differences in grain size. The fine grain P/M alloy was less rough and tortuous than the I/M alloy, resulting in less closure. Although closure was the dominant mechanism causing retardation at low values of ΔK.

Introduction

Fatigue crack growth (FCG) rates for advanced powder metallurgy (P/M) aluminum alloys for aerospace have been shown to be significantly faster than for conventional ingot metallurgy (I/M) aluminum alloys, when tested at low stress intensities under constant amplitude loading (1-3). However, under more realistic aircraft fighter spectrum loading conditions, FCG lives of P/M alloys have been reported to be as much as 50% longer than I/M alloys tested under similar conditions (4-6). The longer FCG lives for P/M alloys over I/M alloys under spectrum loading occurred at high applied stress intensities and appears to be a result of much greater crack growth retardation at high peak overload stress intensities. The FCG rates for P/M alloys at low to moderate stress intensities under spectrum loading, however, are significantly faster than I/M alloys as is usually the case under constant amplitude loading. Bretz (5), however, found that by applying periodic tensile overloads superimposed on a constant amplitude baseline, P/M aluminum alloy 7091-T7E69 exhibits remarkably slower FCG rates compared to I/M aluminum alloy 7075-T7351, when tested at significantly lower stress intensities.

It has been well documented (1-3) that the much finer grain size of P/M alloys which is approximately 5µm, accounts for the faster FCG rates when compared to I/M alloys having a coarse grain size which is approximately 50µm. Increasing grain size of P/M aluminum alloys through thermo-mechanical processing has been shown to have a pronounced effect on decreasing FCG rates at low stress intensities (3, 7-9). Microstructural features such as strengthening precipitates have been shown to affect FCG rates in both P/M and I/M aluminum alloys under constant amplitude loading (7,10,11). Effects of precipitates and constituent particles have also been shown to affect FCG rates in I/M alloys under both constant amplitude and spectrum loading (11-13). It is still unclear, however, why the FCG rates of fine grained P/M alloys are slightly slower than coarse grained I/M alloys under spectrum loading conditions at high stress intensities. The objective of this investigation is to evaluate the effects of tensile overloads and microstructure on the FCG behavior of an advanced experimental P/M aluminum alloy CW67 and to compare this behavior to a conventional I/M aluminum alloy 7475.

Background

Design of aircraft structures using a damage tolerant philosophy uses FCG data generated from constant amplitude loading tests. These data cannot be used reliably to predict the life of a structure under actual spectrum loading. Load fluctuations which are spectral in nature produce periodic overloads which lead to crack growth retardation and longer lives than under equivalent constant amplitude loading. As a result, spectrum fatigue testing is increasingly being used for material selection and design (14). From laboratory tests involving high tensile overloads superimposed on a constant amplitude loading baseline, it has been observed (11,13) that crack growth rates for I/M aluminum alloys are severely retarded extending FCG life up to 20 times compared to constant amplitude loading.

Numerous investigators (15-22) have attempted to explain the mechanisms of crack growth retardation following tensile overloads. It has been suggested that crack growth retardation is caused by residual compressive stresses acting on the crack tip (15). These compressive stresses are believed to result from elastic unloading of the specimen which clamps the overload plastic zone to produce a crack tip compressive stress field.

Elber (18) suggested that permanent tensile displacements are generated in the crack plane due to plastic deformation from an overload. This increased plasticity results in contact between the fracture surfaces in the wake of the advancing crack front even at tensile loads. As a result, the crack remains closed during a portion of the loading cycle. Since the crack cannot propagate while it remains closed, the effective stress intensity range (ΔK_{eff}) responsible for crack growth during cyclic loading is given by

$$\Delta K_{eff} = K_{max} - K_{op} \tag{1}$$

where K_{max} and K_{op} are the maximum and crack tip opening stress intensities, respectively. Such plasticity-induced crack closure arguments have been used to explain the transient effects following single overloads (19). Extensive fractographic studies of overload effects in 2024-T3 aluminum (19) typically showed pre-overload crack advance by a striation mechanism proceeded by a stretch zone due to the overload and subsequently followed by a highly abraded post-overload zone with irregular surface features and no discernable striations. Based on Elber's crack closure arguments, Von Euw (20) suggested that the enhanced residual plastic deformation arising from the overload results in crack face contact in the wake of the advancing crack tip which accounts for crack growth retardation.

Information on crack growth retardation from a number of studies (13,21) reveals that such plasticity-induced crack closure processes cannot fully explain the transient effects due to load excursions. An examination of crack tip profiles following overloads (13,21) indicates that an overload causes severe crack branching, which results in pronounced reductions in the driving force. Such reductions in ΔK_{eff} in the post-overload zone resulting from crack branching and crack closure, may result in an effective driving force comparable to threshold stress intensities, even though moderate baseline stress intensities are applied externally. Thus, branching following an overload suggests that plasticity-induced crack closure in the wake of the crack does not appear to be the only mechanism for retardation following overloads.

Although the plasticity-induced closure concept has been widely used to rationalize retardation effects following overloads, extensive experimental evidence (22) indicates that such a model cannot completely account for all the transient effects observed during post-overload cycling. It is widely acknowledged that Stage II crack growth occurs by a striation mechanism induced by concurrent or alternating slip systems (23). Stage I (near-threshold) crack advance is known to take place primarily along a single slip system (23,24) since the plastic zone size generated by the local driving force is typically smaller than the grain size. Experimental evidence based on crack path observations and fractography in steels (25) indicates that such a Stage I propagation mechanism results in serrated or faceted fracture features and is accompanied by strong Mode II (shear) displacements, even though Mode I (tensile) conditions prevail. The occurrence of such shear displacements during post-overload cycling in the threshold regime has been verified experimentally from crack profiles (21,24,26). The presence of shear displacements and microscopically "rough" fracture surface features readily provides a micro-roughness mechanism (22) for a reduction in ΔK_{eff} due to contact of the fracture surfaces. This roughness-induced closure model provides a mechanistic basis for sustained retardation once Stage I fatigue mechanisms are activated in the post-overload zone, in addition to crack branching and plasticity-induced crack closure in the wake of the advancing crack front.

The mechanisms of crack closure and crack branching on the FCG behavior of aluminum alloys are strongly influenced by their microstructural features such as grain size (1-3,7-9,27), strengthening precipitates (7,10-13,28-30) and constituent particles (11-13). The improved FCG resistance in metals with increasing grain size has been explained by the concepts of crack tip plasticity (27,31-33), reverse slip (27,33) and the characteristics of deformation and fracture (27-36). When the stress intensity is low, the plastic zone size generated at the crack tip is smaller than the grain size in coarse grain materials such as I/M alloys, as shown in Figure 6. This condition results in slip to generally occur on a single slip plane approximately 45° to the tensile axis. Since the deformation is confined within the grain, slip can reverse on this plane upon unloading thereby reducing crack advance. The fracture appearance is highly faceted and crystallographic in nature leading to roughness-induced closure (24-26,37) due to accompanying shear displacements from this highly angular deformation. In fine grain material such as P/M alloys, the plastic zone size at a given stress intensity is larger than the grain size resulting in multiple slip, within the grain in addition to slip in neighboring grains. There is less reversible slip and the grain boundary becomes a less effective barrier to slip. The deformation is less angular resulting in a more planar fracture appearance.

Crack deflection (deviation from a straight line path) is also greater for coarse grain materials due to an increase in slip length resulting in a longer crack path (36). Crack deflection and branching generally occur in coarse grain materials at low stress intensities due to angular deformation resulting from a single slip system being activated. It has been shown with quantitative models (38,39) that crack deflection and its associated branching can significantly reduce the effective stress intensity. In fine grain materials there is very little crack deflection and branching due to multiple slip occurring within several grains resulting in planar deformation and a smooth fracture surface. The increased deflection and branching associated with coarse grain materials results in a rough fracture surface thereby increasing closure (24-26,37).

The effect of precipitates and constituent particles on the FCG behavior of 7XXX aluminum alloys has also been investigated under constant amplitude and spectrum loading (11-13). It has been determined that lower purity I/M alloy 7075 having a higher volume fraction of coarse constituent particles is more resistant to FCG under spectrum loading than higher purity I/M alloy 7475. It has also been observed that high tensile overloads causes secondary cracking at coarse constituent particles resulting in the crack tip stress intensity to be distributed among the local constituent particle fractures reducing the crack tip strain energy, thereby lowering the effective stress intensity (11). Under constant amplitude loading, however, the higher purity I/M alloy 7475 has been found to be more fatigue resistant especially at high stress intensities due to its higher fracture toughness compared to the lower purity I/M alloy 7075 (12). It has also been observed that alloys with higher strength T6 tempers are more resistant to FCG under spectrum loading than lower strength T7 tempers. The higher strength T6 temper promotes more constituent particle fracture than the T7 temper due to more shearing of precipitates. This results in shear bands impinging on hard constituent particles, thereby increasing the stress concentration which ultimately fractures the particles reducing the effective stress intensity. In addition, the T6 temper has been found to have a higher threshold stress intensity than the T7 temper under constant amplitude loading due to shearable precipitates of the T6 temper (11). This would also explain why the T6 temper was found to be more resistant to crack growth than the T7 temper following tensile overloads, since an overload can reduce the applied stress intensity to near threshold stress intensities resulting in slip on a single slip plane.

In rapidly solidified P/M aluminum alloys with little or no constituent particles (40), the decrease in stress intensity due to constituent particle fracture and hence FCG retardation following overloads should not be a factor as in I/M alloys. In addition, the much finer grain size in P/M alloys compared to I/M alloys should limit the amount of reversible slip at relatively low stress intensities, since the plastic zone size should be much larger than the P/M alloy grain size. Furthermore, the precipitate size of P/M alloys in the T7X1 temper is generally much finer than the T73 temper of I/M alloys since P/M alloys are only slightly overaged. It is speculated that the FCG resistance may be enhanced for P/M alloys under spectrum loading due to their inherent higher strength and toughness compared to I/M alloys.

Materials and Experiments

The materials tested in this work were a rectangular extrusion (11.4 cm wide by 3.8 cm thick) of P/M aluminum alloy CW67-T7X1 and a rolled plate (1.9 cm thick) of I/M 7475 aluminum alloy in both the T7351 and T651 conditions. P/M CW67 extrusion was chosen because it is the most advanced high strength P/M aluminum alloy and may directly replace I/M 7475 plate in aerospace applications. In addition, P/M aluminum alloy extrusions generally have enhanced toughness and FCG resistance compared to forgings due to a more homogeneous microstructure resulting from a higher degree of hot working (39). I/M 7475 plate was chosen for comparison, as opposed to an extrusion, due to its current wide use as an advanced structural airframe material. Both alloys contain a relatively low volume fraction of constituent particles. The constituent particles in P/M CW67, however, are much smaller than in I/M 7475 due to rapid solidification.

P/M CW67 was received in the slightly overaged (T7X1) condition for an optimum combination of strength and toughness (41), whereas I/M 7475 was received in the fully overaged stress corrosion resistant (T73) temper. The yield strength of P/M CW67-T7X1, is approximately 30% higher than I/M 7475-T7351, while I/M 7475-T651 has a slightly lower strength than P/M CW67 as shown in Fig. 1. The nominal chemical compositions of P/M aluminum alloy CW67 (41) and I/M aluminum alloy 7475 are given in Table I.

TABLE I - Nominal Chemical Compositions (wt %)

Alloy	Zn	Mg	Cu	Zr	Ni	Cr
P/M CW67	9.0	2.5	1.5	0.14	0.1	-
I/M 7475	5.7	2.2	1.6	-	-	0.22

The major difference in composition is that P/M CW67 contains a significant increase in Zn over I/M 7475. This would result in a higher volume fraction of η' (Zn_2Mg) precipitates upon aging resulting in increased strength. The Mg and Cu contents of P/M CW67 were not noticeably increased compared to I/M 7475 due to their less potent effect on precipitation strengthening. They are added mainly for solid solution strengthening, however, Mg forms with Zn and Cu for precipitation strengthening. The dispersoid forming elements such as Zr and Ni in P/M CW67 are used to inhibit recrystallization during subsequent hot working instead of Cr in I/M 7475. Both alloys are high in purity containing less than 0.2 wt % of iron plus silicon, which was determined from a wet chemical analysis. In addition, CW67 contains approximately 0.35 wt % O_2 (41) resulting from the formation of an oxide film on the surface of the original gas atomized powder particles (39).

31

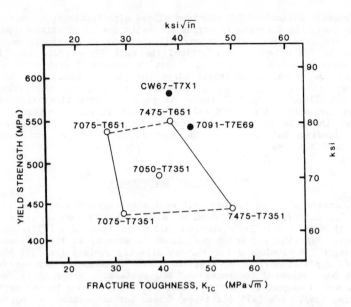

Figure 1 – Yield Strength Versus Fracture Toughness of P/M and I/M 7XXX Aluminum Alloys.

The longitudinal tensile properties and plane strain fracture toughness (K_{IC}) values from the L–T orientation for P/M CW67-T7X1 and I/M 7475-T7351 are presented in Table II (42).

TABLE II – Mechanical Properties

Aluminum Alloy	Yield Strength MPa (ksi)	Tensile Strength MPa (ksi)	Elonga- tion %	Fracture (K_{IC}) Toughness MPa√m (ksi√in)
P/M CW67-T7X1	586 (85)	614 (89)	11	41 (37)
I/M 7475-T7351	448 (65)	510 (74)	12	56 (51)

FCG compact tension specimens with W = 40 mm (6.4 mm thick) were machined in the L–T orientation for both constant amplitude loading (H/W = 0.5) and periodic overloads (H/W = 0.6) for the baseline tests at a stress ratio of 0.33. The remaining tests at R=0.1 were conducted using specimen with w = 40 mm, H/W = 0.6, and a thickness of 0.5 mm.

Fatigue crack growth testing was conducted in accordance with ASTM E647, Standard Test Method for Constant-Load-Amplitude FCG. FCG specimens were precracked under decreasing ΔK to a initial crack length of 1.27 cm (0.5 in.). For the baseline tests at R = 0.33, the specimens were cycled to failure using a maximum load of 66 N (300 lb). For the periodic tensile overload tests, an overload equal to 1.8 times the maximum constant amplitude baseline load was applied every 8,000 cycles. The tests were carried out on a computer-controlled servo-hydraulic MTS machine in laboratory air at room temperature. The cyclic frequency (f) was 25 Hz. Changes in crack length were determined from strain-based compliance measurements (43). The crack growth rates were calculated using a polynomial method described in ASTM E647.

A second series of tests were conducted at R = 0.1 using P/M alloy CW67-T7X1 along with I/M 7475-T651 which has a similar strength. Both materials were subjected to constant amplitude fatigue as well as spectrum loading using an overload of 1.8 every 8000 cycles. In addition, a series of constant ΔK tests were conducted using various values of ΔK. In each test a single overload was applied after steady-state growth rates were achieved. The crack growth rate was monitored subsequent to the overload until a steady-state growth rate was again achieved.

Closure loads were determined from load-strain traces obtained periodically over a complete load-unload cycle using a strain gage that was mounted on the backface of the specimen. The closure loads were determined from the load-strain curves as the point where the curve deviates from linearity upon unloading from maximum load. The closure stress intensity (K_{cl}) was then calculated from the stress-intensity solution for the CT specimen and the effective stress intensity range (ΔK_{eff}) was then determined using eq (1). Crack closure was defined by

$$\text{Crack Closure} = 1 - \frac{\Delta Keff}{\Delta Kapp} \qquad (2)$$

Following FCG testing the specimen fracture surfaces were examined using scanning electron microscopy. Fracture profiles of the FCG specimens were examined microscopically in order to better understand the prominent FCG mechanisms.

Results and Discussion

Optical micrographs of P/M CW67-T7X1 in the extruded condition and I/M 7475-T7351 are shown in Figure 2. Note the difference in magnification for the two materials. The microstructure of CW67 from a transverse section of the extrusion contains equiaxed grains which are approximately 5μm in diameter and are parallel with the extrusion direction. In contrast, the microstructure of I/M 7475-T7351 plate contains large layered grains which are approximately 50μm in thickness. In addition, 7475 contains a small amount of constituent particles, whereas in the refined microstructure of CW67 there are none visible at this magnification. However, at higher magnifications using a TEM, fine constituent particles containing iron and silicon were observed in some areas in CW67. Further, the microstructure of CW67 was found to contain stringers of oxide particles parallel to the extrusion direction.

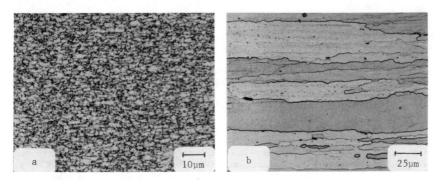

Figure 2 - Optical micrographs of (a) P/M CW67-T7X1 extrusion and (b) I/M 7475-T7351 plate (Note: difference in magnification).

The major difference in microstructure between P/M CW67 and I/M 7475 is the grain size, which is approximately an order of magnitude smaller in P/M CW67. The precipitate size is somewhat finer in P/M CW67 enabling the alloy to attain a high strength without the concern for SCC resistance as in I/M 7475. Constituent particles in P/M CW67 are not visible in the optical micrographs as they are in I/M 7475.

Crack length versus number of cycles is presented in Fig. 3 for P/M CW67 and I/M 7475 at R = 0.33 for both constant amplitude (CA) and periodic overload (PO) conditions. Note that a log scale is used for the number of cycles in order to compare the CA FCG lives with the PO lives. Under CA loading, the FCG life of P/M CW67 is significantly shorter than I/M 7475. The fatigue crack for both materials grew approximately 15 mm to failure from an initial crack length of 13 mm.

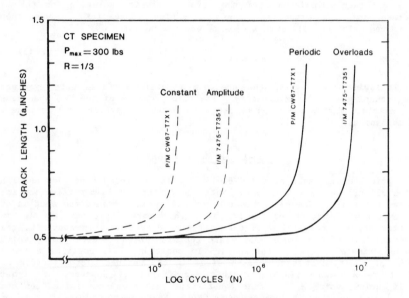

Figure 3 - Crack length versus FCG life of P/M CW67-T7X1 and I/M 7475-T7351 under variable loading conditions.

Under PO loading, the FCG life of P/M CW67 is again significantly shorter than I/M 7475. The fatigue crack for both materials grew approximately 20 mm to failure. When the alloys experience an 80% tensile overload applied every 8,000 cycles, the FCG life for both materials is seen to dramatically increase approximately 17 times over CA loading. The difference in FCG life between P/M CW67 and I/M 7475 stays relatively the same under both loading conditions. P/M CW67 exhibits approximately a 3 fold shorter FCG life than I/M 7475 under both loading conditions.

The data of Fig. 3 are replotted as da/dN against ΔK in Fig. 4. It can be seen that under CA loading, P/M CW67, exhibits a 2 to 3 times faster FCG rate than I/M 7475 throughout the stress intensity range investigated. Under PO loading, P/M CW67 again exhibits a 2 to 3 times faster FCG rate than I/M 7475 at low to moderate stress intensities (ΔK = 4-6 MPa$\sqrt{m}$). At higher stress intensities (ΔK = 10 MPa$\sqrt{m}$), however, the FCG rate of P/M CW67 approaches the rate for I/M 7475 and it appears that a crossover takes place with further increases in stress intensity. In addition, the slope of the FCG rate curve of P/M CW67 above a moderate stress intensity level of 6 MPa√m appears to decrease. This increased FCG retardation for P/M CW67 at moderate to high stress intensities will be explained in more detail later when analyzing the fracture behavior. Overall, the FCG rate for both alloys is approximately 15 times slower at higher stress intensities under PO loading when compared to CA loading.

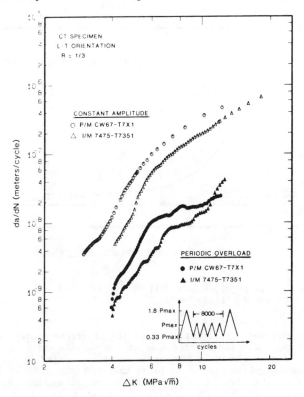

Figure 4 - FCG Rate Data Versus Applied Baseline Stress Intensity Range (ΔK) for P/M CW67 and I/M 7475 under Variable Loading Conditions.

Results obtained from closure measurements are presented in Fig. 5. The data from tests at R = 0.1 is included on the P/M alloy which are part of the second series of tests to be discussed later. Because of the fairly large scatter in the closure data and the difficulty in defining the closure load precisely, only trend lines are shown in Fig. 5. Plotted is the crack closure quantity defined in equation (2) which is one minus the ratio of the effective stress intensity range to the applied stress intensity range. When the effective stress intensity equals the applied stress intensity, there is no closure present. At R = 0.33, the amount of closure for P/M CW67 was zero throughout the applied stress intensity range. However, for I/M 7475, there is a significant amount of closure at low stress intensities. With increasing stress intensity range, the amount of closure exhibited by I/M 7475 gradually decreases close to zero. The effect of decreasing the mean stress by changing R to 0.1 produced a significant increase in the amount of closure for P/M CW67 especially at lower stress intensities. At R = 0.33, the differences in closure between I/M 7475 and P/M CW67 may explain the differences in growth rates at low values of ΔK. At high values of ΔK (greater than $\Delta K = 6$ MPa$\sqrt{m}$), however, differences in closure are not apparent and, therefore, cannot be used to explain the differences in growth rates between the two materials.

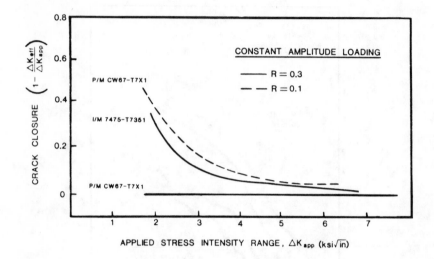

Figure 5 - Degree of closure with ΔK for P/M CW67 and I/M 7475 under constant amplitude loading.

Fracture surfaces were observed using SEM to relate crack growth rates to fatigue mechanisms. SEM fractographs at low magnification are presented for CA loading for a relatively low stress intensity range ($\Delta K = 5$ MPa$\sqrt{m}$) in Fig. 6. At this magnification it is clear that P/M CW67 has a much smoother fracture surface than I/M 7475. At a higher magnification, Fig. 7, the fracture behavior of P/M CW67 (Fig. 7a) appears to be predominantly transgranular and uniform throughout the surface of the specimen. The fracture surface for I/M 7475 (Fig. 7b) appears to be highly faceted and predominantly transgranular. The size of the faceted fracture features appears to correspond with the grain size. In addition, secondary cracking is observed on the fracture surface. Crack profiles sectioned from the specimen midplane are shown in Fig. 8. Fig. 8a for P/M CW67 illustrates the transgranular fracture behavior. In addition, the size of the fracture features on the surface in Fig. 7a appears to correspond with the grain size.

For I/M 7475, Fig. 8b clearly shows evidence of secondary cracking. Secondary cracking is attributed to the deflection of the crack path, which is known to result in I/M alloys with large grain sizes that exhibit planar slip (38). Thus, the large grains in I/M 7475 appear to strongly influence the fracture behavior by deflecting the crack path at grain boundaries. This, in turn, produces a reduction in the effective crack driving force. Larger grains producing a longer more deflected crack path and secondary cracking has been found to significantly enhance the FCG resistance of other aluminum alloys (9,34,35,38).

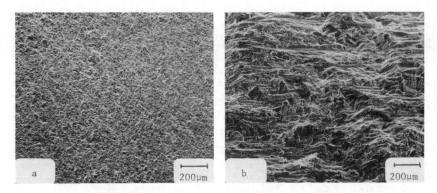

Figure 6 – SEM fractographs at low magnification of (a) P/M CW67 showing a much smoother fracture surface than (b) I/M 7475 ($\Delta K = 5$ MPa√m).

FCG →

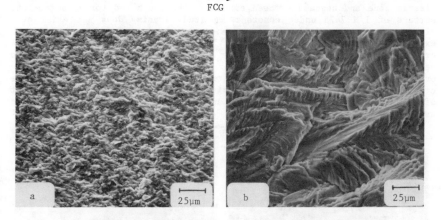

Figure 7 – SEM fractographs at high magnification of (a) P/M CW67 showing a more uniform and less tortuous fracture surface than (b) I/M 7475 ($\Delta K = 5$ MPa√m).

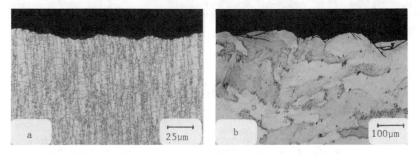

Figure 8 – Crack profiles of (a) P/M CW67 showing less crack deflection and no secondary cracking as in (b) I/M 7475 ($\Delta K = 5$ MPa√m) (Note: difference in magnification).

Similar analyses were made of the fracture surfaces of the specimens subjected to PO loading. The fracture surface of P/M CW67 is shown in Fig. 9a at a relatively low baseline stress intensity (ΔK = 5 MPa$\sqrt{m}$). The overloads are clearly marked in addition to the post-overload fracture surface, which appears to extend approximately 10μm from the application of the overload. The post-overload fracture surface is relatively flat following the overload and is presumed to result from compressive stresses in the wake of the crack tip due to an enlarged overload plastic zone. The fracture surface following the post-overload fracture surface returns to the transgranular fracture appearance as previously shown in Fig. 7a under constant amplitude loading.

The fracture surface of I/M 7475 following tensile overloads is shown in Fig. 9b at the same low stress intensity (ΔK = 5 MPa$\sqrt{m}$). An obvious observation in this fractograph is that the overload lines are more closely spaced compared to the P/M CW67 fractograph indicating a slower FCG rate for I/M 7475. The fracture surface of I/M 7475 following an overload is very flat compared to the tortuous fracture surface under constant amplitude loading, shown in Fig. 7b. This is presumably due to extensive compressive stresses in the wake of the crack tip resulting from an enlarged overload plastic zone and hence increased closure. The rough and tortuous fracture surface of I/M 7475 under constant amplitude loading does not exist even after many cycles following a tensile overload presumably due to an extensive amount of crack closure. In addition, there is evidence of cracking along the overload line for I/M 7475 which appears to be more severe than the cracking observed for P/M CW67. This cracking along the overload line could also contribute to reduced stress intensities at the crack tip, thereby increasing the resistance for crack growth. The extent of this cracking into the bulk of the specimen for both alloys was not found to be severe.

FCG

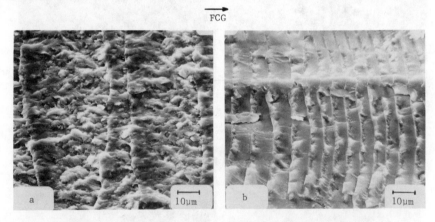

Figure 9 - SEM fractographs showing the effect of tensile overloads on the fracture behavior for (a) P/M CW67 and (b) I/M 7475 (ΔK = 5 MPa$\sqrt{m}$).

The fracture surface of P/M CW67 at a higher baseline stress intensity range (ΔK = 15 MPa$\sqrt{m}$) is shown in Fig. 10a. The cracking along the overload line seems more severe than at lower baseline stress intensities. In the post-overload zone, the fracture surface is again relatively flat. The cracking does not significantly penetrate below the fracture surface as shown in the crack profile (Fig. 11a) indicating that this cracking probably does not significantly affect the FCG resistance.

FCG →

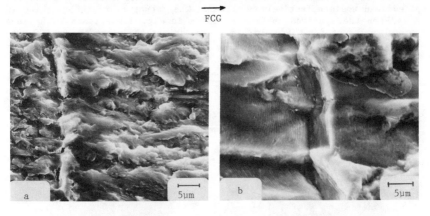

Figure 10 - SEM fractographs showing the effect of tensile overloads on the fracture behavior for (a) P/M CW67 and (b) I/M 7475 (ΔK = 15 MPa$\sqrt{m}$).

The fracture surface of I/M 7475 at the same high stress intensity range as P/M CW67 is shown in Fig. 10b. Extensive cracking along the overload line is observed which significantly penetrates below the fracture surface as shown in the crack profile (Fig. 11b). It is interesting to observe that crack growth in the pre-overload region has occurred by the mechanism which produces striations on the fracture surface. However, in the post-overload region, where plasticity-induced crack closure is presumed to occur, the striations appear to have been obliterated probably from compressive stresses. In addition, it appears that abrasion of the fracture surface has occurred in the post-overload zone presumably due to fracture surface interference resulting from crack closure. Overall, the more extensive cracking along the overload line for I/M 7475 compared to P/M CW67 at higher baseline stress intensities is presumed to enhance the FCG resistance. However, the slope of FCG rate curve for P/M CW67 at higher baseline stress intensities appears to decrease more I/M 7475 as previously shown in Fig. 4. The improved FCG resistance of P/M CW67 at higher stress

← FCG

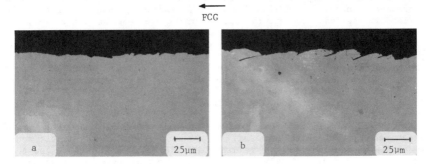

Figure 11 - Crack profiles showing the extend of branched cracks from overloads for (a) P/M CW67 and (b) I/M 7475 (ΔK = 15 MPa$\sqrt{m}$).

intensities over that of than I/M 7475 maybe due to extensive crack branching on the specimen surface resulting from overloads, as shown in Fig. 12a. The crack profile near the edge appears rough and tortuous which may result in crack wedging, thereby reducing the stress intensity since the applied load would effectively be lower. The amount of crack branching in I/M 7475 on the specimen surface as shown in Fig. 12b appears to be less severe than for of P/M CW67, which may explain why the slope of the FCG rate curve for I/M 7475 does not decrease as much at higher baseline stress intensities.

FCG

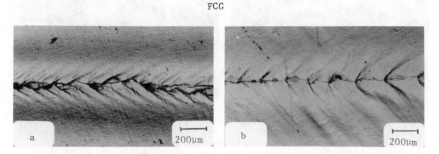

Figure 12 - Crack profiles showing the extent of branched cracks on the specimen surface from overloads for (a) P/M CW67 and (b) I/M 7475 ($\Delta K = 15$ MPa$\sqrt{m}$).

The results of the CA and PO loading tests on P/M CW67 and I/M 7475 are summarized in Fig. 13 for tests conducted at R=0.1. Recall that for this series of tests, the I/M alloy was 7475-T651 was also used which has a yield strength very similar to that of the P/M alloy (see Fig. 1). At this value of R=0.1, the PO test results for P/M CW67 do not lie very far below those for CA loading. In fact, the net retardation due to overloads results in growth rates approximately a factor of two slower than under CA loading whereas the retardation was generally close to a factor of ten for CW67 when the tests were conducted at R=0.33. The retardation due to overloads in I/M 7475, on the other hand, is similar at R=0.1 to that observed at R=0.33. In both cases, periodic overloads retard the growth rate by approximately a factor of ten.

To further evaluate the retardation due to single overloads, a series of tests were conducted under constant ΔK conditions. A typical set of data is presented in Fig. 14 which illustrates a constant growth rate before the overload is applied. The growth rate after the overload is much slower, but gradually increases until the steady state growth rate is again achieved. Defined in Fig. 14 are the terms delay cycles and delay distance. The delay cycles are the net gain in cycles due to the application of the overload, ie. the offset in cycles between the steady state lines before and after overload. The delay distance is the amount of crack extension after an applied overload before steady-state growth is again achieved. From the results of a series of tests at various constant values of ΔK, data were obtained for these two quantities. The results for delay cycles are presented in Fig. 15 and the results for delay distance, normalized with respect to the calculated size of the plane stress overload plastic zone, are presented in Fig. 16. The results for delay cycles, Fig. 15, show that for all values of ΔK I/M 7475 alloy demonstrates more crack retardation due to an overload than P/M CW67 under identical conditions. It can also be seen that the number of delay cycles decreases with increasing ΔK except at very high ΔK where the number of delay cycles starts to increase for I/M 7475. Noting that the PO loading spectrum consisted of an overload every 8000 cycles, it can be seen from Fig. 15 that except at very low values of ΔK, the periodic overloads do not cause a cumulative effect because steady state growth rates are achieved in less than 8000 cycles. This occurs for ΔK values greater than approximately 6 MPa√m for I/M 7475 and 5 MPa√m for P/M CW67. These data also corroborate the relatively small retardation effect in PO loading compared to CA loading for P/M CW67 as shown in Fig. 13. The data for I/M 7475 also indicate that large differences between PO and CA loading should not be expected since the number of delay cycles for each overload (applied at 8000 cycle intervals) is typically of the order of several thousand cycles. The data of Fig. 13, however, show a much larger retardation effect. One possible explanation is that the data of Fig. 15 were obtained under <u>constant</u> ΔK conditions whereas the spectrum load data of Fig. 13 were obtained under constant load, <u>increasing</u> ΔK conditions. Development of closure under increasing ΔK has been shown to be different than that under constant ΔK in some materials (44) and may account for the apparent anomaly in the data.

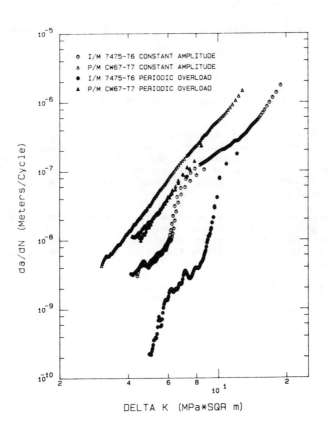

Figure 13 - FCG rate data for P/M CW67 - T7X1 and I/M 7475 - T651 at R = 0.1.

41

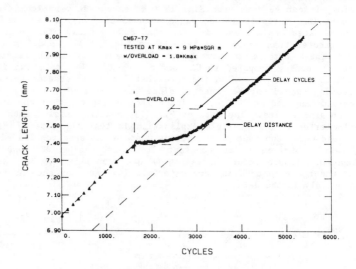

Figure 14 - Effect of a tensile overload on the
number of delay cycles for P/M CW67.

The normalized delay distance, shown in Fig. 16, is seen to be approximately one for both materials over the entire range of ΔK. This means that crack growth retardation due to an overload occurs until the crack grows out of the overload plastic zone. This concept, which is widely used in crack growth retardation modeling, is validated by the present set of experimental data.

Finally, crack closure measurements were made under constant ΔK conditions periodically after the overload was applied. A typical set of load-differential strain curves taken after an applied overload, are presented in Fig. 17. The differential strain is the difference between the actual load-strain data and a straight-line fit to the linear region at high loads. Amplification of these differential data allow for easy determination of closure loads. The data presented are for I/M 7475 at a constant ΔK of 8 MPa$\sqrt{m}$ where the number of delay cycles was approximately 3000. It can be seen that the closure curves develop progressively from a bilinear form to a trilinear form as the number of cycles after the overload increases. From these curves, it was not apparent where the closure load should be chosen, ie. at the lower or upper deviation from linearity. Further, at best, the closure load is either constant or an increasing function with cycles. In either case, the closure data do not appear to be able to explain the gradual increase in growth rate following an overload as shown in Fig. 14. One would expect to find high closure loads (small effective ΔK) immediately after an overload followed by a gradual decrease in closure (increase in effective ΔK) with number of cycles. Clearly the data of Fig. 17 show no such trend. For this reason, and because of the lack of an unambiguous closure load from the data, the closure data obtained under constant ΔK conditions were not used further in the analysis.

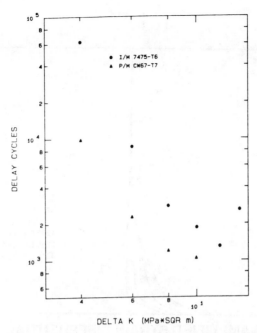

Figure 15 - Effect of stress intensity on the number of delay cycles following tensile overloads.

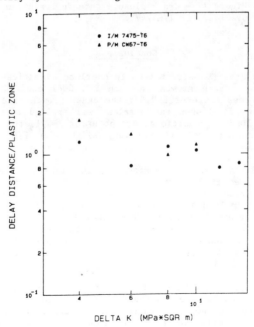

Figure 16 - Effect of stress intensity on the delay distance following overloads.

43

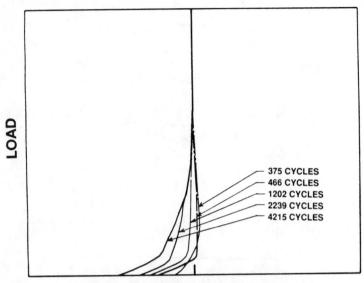

LOAD

375 CYCLES
466 CYCLES
1202 CYCLES
2239 CYCLES
4215 CYCLES

AMPLIFIED STRAIN DIFFERENTIAL

Figure 17 - Crack closure data following an
overload for I/M 7475.

Conclusions

Application of a 80 percent tensile overload significantly retards the
constant amplitude crack growth in both P/M CW67 and I/M 7475 aluminum
alloys. At a stress ratio of R=0.33, the crack growth rates were retarded
by over a factor of 10 when the overload was applied every 8000 cycles.
Under identical loading conditions, except using R=0.1, the net retardation
due to this spectrum overload was very slight for P/M CW67 while the
retardation in I/M 7475 was again approximately a factor of 10.

Analysis of the fracture surface in the region directly after the
overload was applied at low stress intensities evidence that crack closure
was the dominant retardation mechanism, while crack branching was judged to
have only a minor influence. At high stress intensities, on the other hand,
crack branching appeared to play a more dominant role. For field closure
measurements using back face strain, there were no significant changes in
closure load even though the characteristics of the load–strain curves after
an applied overload were noticeably different from those observed under
constant amplitude loading.

Overall, the FCG resistance of P/M CW67 under both constant amplitude
and periodic overload loading was found to be inferior to that of I/M 7475.
Closure measurements as well as fractography showed that the differences in
growth rate were due primarily to differences in closure in the two alloys.
The fracture surface of P/M CW67 which has a much finer grain size, was less
rough and tortuous than for I/M 7475 resulting in less closure.

References

1. R. E. Sanders, Jr., W. L. Otto and R. J. Bucci, "Fatigue Resistant Aluminum P/M Alloy Development" (AFML-TR-79-4131, Wright-Patterson AFB, OH, September 1979).

2. D. P. Voss, "Structure and Mechanical Properties of Powder Metallurgy 2024 and 7075 Aluminum Alloys" (Final Report EOARD-TR-80-1, European Office of Aerospace Research and Development, October 1979).

3. V. W. C. Kuo and E. A. Starke, Jr., "The Effect of ITMT's and P/M Processing on the Microstructure and Mechanical Properties of the X7091 Alloy," Met. Trans. 14A (1983) 435-447.

4. J. J. Ruschau, "Spectrum Fatigue Crack Growth Rate Characteristics of PM Aluminums 7090 and 7091" (AFWAL-TR-83-4032, Wright-Patterson AFB, OH, April 1983).

5. P. E. Bretz, Spectrum Fatigue Behavior, An Update (pamphlet), (Alcoa Tech. Brief., March 1984).

6. G. V. Scarich and K. M. Bresnahan, "Spectrum Fatigue Crack Growth Behavior of High Strength 7XXX RST Aluminum Alloys" (Paper presented at the TMS Fall Meeting, Toronto, Canada, October 1985).

7. Y. W. Kim and L. R. Bidwell, "Effects of Microstructure and Aging Treatment on the Fatigue Crack Growth Behavior of High Strength P/M Aluminum Alloy X7091," High-Strength Powder Metallurgy Aluminum Alloys, TMS (1982) 107-124.

8. Y. W. Kim and W. M. Griffith, "The Effect of Grain Size on Strength and Fatigue Crack Growth Behavior in P/M Aluminum Alloy 7091," (Proceedings of the P/M Aerospace Materials Conference, Berne, Switzerland, November 1984).

9. P. E. Bretz, J. I. Petit and A. K. Vasadevan, "The Effects of Grain Size and Stress Ratio on Fatigue Crack Growth in 7091 Aluminum Alloy," Fatigue Crack Growth Threshold Concepts, TMS (1984) 163-184.

10. S. Suresh and A. K. Vasudevan, "Application of Fatigue Threshold Concepts to Variable Amplitude Crack Propagation," Fatigue Crack Growth Threshold Concepts, TMS (1984) 361-378.

11. P. E. Bretz, A. K. Vasudevan, R. J. Bucci and R. C. Malcolm, "Fatigue Crack Growth Behavior of 7XXX Aluminum Alloys under Simple Variable Amplitude Loading," ASTM STP 833 (1984) 242-265.

12. T. H. Sanders, Jr. and J. T. Staley, "Review of Fatigue and Fracture Research on High-Strength Aluminum Alloys," Fatigue and Microstructure, ASM (1979) 467-522.

13. R. J. Bucci, A. B. Thakker, T. H. Sanders, R. R. Sawtell and J. T. Staley, "Ranking 7XXX Aluminum Alloy Fatigue Crack Growth Resistance Under Constant Amplitude and Spectrum Loading," ASTM STP 714 (1980) 41-78.

14. G. R. Chanani, G. V. Scarich, P. E. Bretz and A. A. Sheinker, "Spectrum Fatigue Crack Growth Behavior of Aluminum Alloys," (Proceedings of the Sixth International Conference on Fracture, New Delhi, India, December 1984).

45

15. C. M. Hudson and H. F. Hardrath, "Effects of Changing Stress Amplitude on the Rate of Fatigue-Crack Propagation in Two Aluminum Alloys," (NASA TN-D-960, September 1961).

16. O. E. Wheeler, "Spectrum Loading and Crack Growth," J. Basic Eng. Trans. ASME 94 (1972) 181-186.

17. J. Willenborg, R. M. Engle and H. A. Wood, "A Crack Growth Retardation Model Using an Effective Stress Concept," (TM-71-1FBR, Wright-Patterson AFB, OH, 1971).

18. W. Elber, "The Significance of Fatigue Crack Closure," ASTM STP 486 (1971) 230-242.

19. E. F. J. von Euw, R. W. Hertzberg, and R. Roberts, "Delay Effects in Fatigue Crack Propagation," ASTM STP 513 (1972) 230-259.

20. E. F. J. von Euw, "Effect of Overload Cycle(s) on Subsequent Fatigue Crack Propagation in 2024-T3 Aluminum Alloy" (Ph.D. Thesis, Lehigh University, 1971).

21. J. Lankford and D. L. Davidson, "The Effect of Overloads Upon Fatigue Crack Tip Opening/Closing Loads in Aluminum Alloys," Advances in Fracture Research vol. 2 (Pergamon Press, Oxford, 1981) 899-906.

22. S. Suresh, "Crack Growth Retardation Due to Micro-Roughness: A Mechanism for Overload Effects in Fatigue," Scripta Met. 16 (1982) 959-999.

23. P. J. E. Forsyth, Proceedings of a Symposium on Crack Propagation, Cranfield, England (1961).

24. R. O. Ritchie and S. Suresh, "Some Consideration on Fatigue Crack Closure Induced by Fracture Surface Morphology," Met. Trans. 13A (1982) 937-940.

25. K. Minakawa and A. J. McEvily, "On Crack Closure in the Near-Threshold Region," Scripta Met. 15 (1981) 633-636.

26. S. Suresh and R. O. Ritchie, "A Geometric Model for Fatigue Crack Closure Induced by Fracture Surface Roughness," Met. Trans. 13A (1982) 1627-1631.

27. J. Lindigkeit, G. Terlinde, A. Gysler, and G. Lutjering, "The Effect of Grain Size on the Fatigue Crack Propagation Behavior of Age-Hardened Alloys in Inert and Corrosive Environment," Acta Met. 27 (November 1979) 1717-1726.

28. T. H. Sanders, Jr., "Factors Influencing Fracture Toughness and Other Properties of Aluminum Lithium Alloy" (Final Report, NADC Contract No. N62269-76-C-0271).

29. J. Lindigkeit, A. Gysler and G. Lutjering, "The Effect of Microstructure on the Fatigue Crack Propagation Behavior of an Al-Zn-Mg-Cu Alloy," Met. Trans. 12A (1981) 1613-1619.

30. S. Suresh, A. K. Vasudevan and P. E. Bretz, "Mechanisms of Slow Fatigue Crack Growth in High Strength Aluminum Alloys: Role of Microstructure and Environment," Met. Trans. 15A (February 1984) 369-379.

31. J. L. Robinson and C. J. Beevers, "The Effect of Load Rates, Interstitial Content, and Grain Size on Low-Stress Fatigue-Crack Propagation in α-Titanium," Met. Sci. Jour. 7 (1973) 153-159.

32. G. R. Yoder, L. A. Cooley and T. W. Crooker, "Observations on the Generality of the Grain Size Effect on the FCG in α+β Titanium Alloys," Titanium '80, TMS (1981) 1865-1873.

33. E. Hornbogen and K. H. Zum Gahr, "Microstructure and Fatigue Crack Growth in a γ-Fe-Ni-Al Alloys," Acta. Met. 24 (1976) 581-592.

34. P. E. Bretz, L. N. Mueller and A. K. Vasudevan, "Fatigue Properties of 2020-T651 Aluminum Alloy," Aluminum-Lithium Alloys II, TMS (1983) 543-559.

35. A. K. Vasudevan, P. E. Bretz, A. C. Miller and S. Suresh, "Fatigue Crack Growth Behavior of Aluminum Alloy 2020," Mat. Sci. Eng. 64 (1984) 113-122.

36. J. Masounave and J. P. Bailon, "Effect of Grain Size on the Threshold Stress Intensity Factor in Fatigue of a Ferritic Steel," Scripta Met. 10 (1976) 165-170.

37. N. Walker and C. J. Beevers, "A Fatigue Crack Closure Mechanism in Titanium," Fatigue Eng. Mat. Struc., 1 (1979) 135-148.

38. S. Suresh, "Crack Deflection: Implications for the Growth of Long and Short Fatigue Cracks," Met. Trans. 14A (November 1983) 2375-2385.

39. H. Kitagawa, R. Yuuki and O. Toshiaki, "Crack-Morphological Aspects in Fracture Mechanics," Eng. Frac. Mech. 7 (1975) 515-529.

40. J. P. Lyle and W. S. Cebulak, "Powder Metallurgy Approach for Control of Microstructure and Properties in High Strength Aluminum Alloys," Met. Trans. 6A (1975) 685-699.

41. G. J. Hildeman, L. C. Labarre, A. Hafeez and L. Angers, "Microstructural, Mechanical and Corrosion Evaluations of a 7XXX P/M Aluminum Alloy (CW67)," High-Strength Powder Metallurgy Aluminum Alloys II, TMS (1986) 25-42.

42. J. J. Kleek, "Effects of Tensile Overloads and Microstructure on the Fatigue Crack Growth Behavior of P/M Aluminum Alloy CW67," (MS Thesis, University of Dayton, 1987).

43. D. C. Maxwell, "Strain Based Compliance Method for Determining Crack Length for a C(T) Specimen," (AFWAL-TR-87-4046, Wright-Patterson AFB, OH, 1987).

44. R. Sunders, "Fatigue Crack Propagation under Stress and K-Controlled Spectrum Loading," Fatigue 84, EMAS (1987) 881-892.

ACOUSTIC INTERROGATION OF FATIGUE OVERLOAD EFFECTS

O. Buck, D. K. Rehbein, and R. B. Thompson

Ames Laboratory
Iowa State University,
Ames, Iowa   50011

Abstract

It is known for some time that fatigue overloads produce an increase of the crack closure stress intensity factor and thus a smaller effective stress intensity range during the crack retardation period. Recently, acoustic techniques have dramatically improved, particularly in spatial resolution as well as in the frequency information. In addition, our interpretation of the received signals in terms of the geometry of asperity contact in the closure region of a fatigue crack has significantly advanced. Thus, it is possible now to characterize the effects of overloads in the wake of the crack in more detail. The present paper summarizes the status of the theory on acoustic transmission and diffraction in the closure region of fatigue cracks. Also reviewed is a new theory in which asperity contact generates a stress intensity factor on the crack tip. This theory uses the acoustically obtained information to deduce information on crack tip shielding. Acoustic experiments performed on fatigue cracks after an overload block clearly show the physical contact of the fracture surfaces produced by the overload. The results obtained so far are discussed.

Introduction

During the growth of a fatigue crack, contact (1,2) between the crack faces is often developed via a variety of mechanisms, including general plastic deformation, sliding of the two faces with respect to one another, or the collection of debris such as oxide particles (3). Figure 1 schematically sketches such a situation. The results of these contacts are twofold. First, compressive stresses are created in the material on either side of the partially contacting crack surfaces. In reaction, an opening load arises which produces a local stress intensity factor, K_I(local), which shields the crack tip from the variations of the externally applied stress intensity factor, K_I(global) (4). This shielding occurs below a stress intensity factor, $K_{Iclosure}$, at which the first contact during unloading occurs. Thus, a first consequence of asperity contact is that the applied stress intensity range, $\Delta K = K_{Imax} - K_{Imin}$, which is usually considered to provide the driving force for fatigue crack propagation may have to be modified (1) to include the effects of crack tip shielding. We have now succeeded (5) in using information from acoustic transmission and diffraction experiments, obtained under plane strain conditions, to determine the size and density of the contacting asperities in the closure region of a fatigue crack, grown under constant ΔK conditions. In this case, we have also succeeded in estimating values for the static stress across a partially closed crack as well as the stress intensity factor, K_I(local), which shields the crack tip below $K_{Iclosure}$. This paper presents a short review of the theory of scattering of acoustic waves, some brief remarks on a "spring model" used to represent the closure region of the crack, and the results obtained on a fatigue crack grown under constant ΔK loading (5-19). This is followed by a discussion of first observations on a fatigue crack grown under constant ΔK conditions except for an overload block. The effects of the overload block on crack closure can be clearly seen in the acoustic transmission and diffraction experiments, even after crack growth has resumed. Although the work is still in progress, the results obtained so far indicate that it is feasible to determine the shielding stress intensity factor after an overload on the crack.

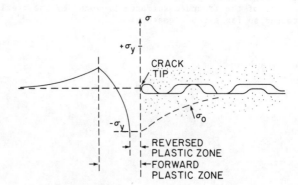

Figure 1 - Schematic illustration of asperity contact near a crack tip. Included is a sketch of the average static stresses across the partially closed crack.

Experimental Procedures

Compact tension specimens of Al 7075-T651 and Al 2024-T351 have been cyclically loaded in an electrohydraulic system in lab air using load shedding until the crack has been initiated. The crack is then grown under constant ΔK_1 loading. To investigate the effects of an overload block, the stress intensity range was increased to $\Delta K_2 = 2\Delta K_1$ for about 20 cycles with subsequent cycling at the earlier level of ΔK_1, which retarded the crack growth for about 10^5 cycles.

The ultrasonic inspection of the fatigue crack was performed in a water tank. For this purpose the starter notches and cracks had to be sealed with silicone caulking to prevent water from entering the crack. The transmitting transducer was aligned such that the acoustic beam interrogates the crack at normal incidence as shown in Fig. 2. A broadband (2-15 MHz) transducer with longitudinal polarization, focused in the plane of the crack was used. This focusing provides the spatial resolution necessary to study the details of the variation of closure, particularly at high frequencies where a spot size of about 1 mm is achieved. Detection is performed with an identical receiving transducer positioned in various orientations. So far we have studied the following arrangements. (a) At $\theta = 0°$, which provides the through transmission signal, (b) At $\theta = 45°$ or $\theta = -45°$ which provides the diffracted signals. The samples were translated with respect to the transducer set-up such that the signals from the unbrokèn ligament (for reference purposes) and the fatigue crack were determined as a function of position x in the plane of the crack. At each of the positions where signal amplitude data were recorded, the received through-transmission waveform was digitized at a 10 nsec sampling rate using a Tektronix digital processing oscilloscope. The data was stored for further processing, such as Fourier analysis, in a Digital Equipment Corporation LSI-11 computer.

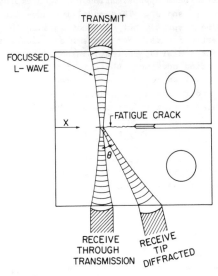

Figure 2 - Experimental geometry for the measurement of through transmission and diffracted signals.

51

Elastic Wave Scattering from Fatigue Cracks

The theory for the scattering of elastic waves from fatigue cracks is based on the electromechanical reciprocity theory of Auld (20) which states that the flaw induced change in the signal transmitted from an illuminating to a receiving transducer, $\delta\Gamma$, is given by

$$\delta\Gamma = \frac{j\omega}{4P} \int_A (u_i^R \tau_{ij}^T - u_i^T \tau_{ij}^R) \, n_j dA \qquad (1)$$

where u_i^R, τ_{ij}^R are the displacement and stress fields that would be produced if the receiving transducer irradiated a flaw free material; u_i^T and τ_{ij}^T are the displacement and stress fields produced when the flaw is irradiated by the transmitting transducer, P is the electrical power incident on the transmitting transducer, and integration is performed over the surface of the scatterer, which has a normal n_j. For the experimental setup, used in the present experiments and schematically shown in Fig. 2, the signal produced by the crack is then given by (18)

$$\Gamma = \frac{j\omega}{4P} \int_{A+} (2u_i^I - \Delta u_i^T) \, \tau_{ij}^R \, n_i^+ \, dA. \qquad (2)$$

where Δu_i^I is the displacement field of the incident illumination, Δu_i^T the crack-opening displacement (COD) in the noncontacting areas of the crack due to the stress field of the acoustic wave and n_i^+ the normal to the crack surface, illuminated by the acoustic wave.

Computation of the crack scattering thus requires three sets of fields to be known. One must know the stress radiation pattern of the receiver, τ_{ij}^R, the displacement radiation pattern of the transmitter, u_i^I, and the dynamic crack-opening displacement, Δu_i^T. As reported previously (18), a scalar Gaussian beam approximation has been employed to estimate the radiation fields τ_{ij}^R and u_i^I. This model includes such effects as diffraction induced beam spread, but does not include the full tensor character of the elastic fields. For beams whose widths are several wavelengths in extent, the scalar approximation should be reasonable since the direction of polarization does not substantially vary over the beam cross-section. In essence, τ_{ij}^R and u_i^I are well characterized quantities for a set of transmitting and receiving transducers. The major problem in evaluating Eq. (2) is in selecting an appropriate description of Δu_i^T.

Spring Model

The simplest model to represent a contacting fatigue crack is a spring model (10), in which the partially contacting interface, in the z=0 plane, is represented by the modified boundary conditions

$$\sigma_{3i}^+ = \sigma_{3i}^- \qquad (3)$$

$$\sigma_{3i}^+ = \kappa_{ij}(u_j^+ - u_j^-) \qquad (4)$$

where the superscripts "+" and "-" refer to the two sides of the interface. The matrix κ_{ij} may be thought of as representing a set of massless springs joining the two sides of the interface. For simple interface topographies, this matrix will be diagonal, with κ_{11} and κ_{22} representing the contact induced resistance to shear and κ_{33} representing the resistance to compression. The properties of the interface are assumed to be linear. This requires that there be resistance to tension as well as compression, which is true when the dynamic stresses of the ultrasonic wave are small with respect to the static stresses associated with the contact. Baik and Thompson (12) have developed a quasi-static model relating κ_{33}, hereafter abbreviated as κ, to solutions of static deformation problems for a variety of crack topographies. κ is found to be a function of both the contact density and dimensions. For sparse, penny shaped contacts, $\kappa = N\pi E'd/8$, where N is the contact density, d is their diameter, $E' = E/(1-\nu^2)$, E = Young's modulus, and ν = Poisson's ratio.

In the spring model, if one assumes normal illumination of the interface by a plane wave having displacement amplitude u_i^I, one finds that for a continuously varying κ the COD is equal to

$$\Delta u_i = \frac{-2j\pi\rho vf/\kappa}{1+j\pi\rho vf/\kappa} u_i^I.$$ (5)

The acoustic signal produced by the crack is then given by

$$\Gamma = \frac{j\omega}{2P} \int_{A^+} [1 + j\alpha]^{-1} u_3^I \tau_{33}^R \, dA$$ (6)

where $\alpha = \pi\rho vf/\kappa$ and the quantity in brackets is equal to the plane wave transmission coefficient. In through transmission, excellent fits of Eq. (6) to experimental data as a function of the measurement frequency are generally obtained when κ is viewed as an adjustable parameter. This agreement has been achieved for a variety of samples ranging from saw slots (simulating the ideal, asperity free condition) to fatigue cracks (9,10,14,15), with one example for a fatigue crack grown under constant ΔK conditions being discussed in the following section.

On the other hand, in using this spring model calculation to predict the diffracted signals from the closure region it has been found that the experimental signals can be considerable greater than the predictions of the model (15). This deficiency has been suggested to be a consequence of the absence of discrete contacts in the model. The above model was expanded (14,15) in an attempt to rectify this problem. Discrete contacts are introduced into the model such that their average κ still represents the spring constant used in Eq. (6). The through transmission results are therefore unchanged. Since κ is a function of N and d, an additional parameter (the contact density N, e.g.) may now be chosen to adjust the calculated to the experimentally observed diffracted signals. Indeed, it has been found that N strongly influences the strength of the diffracted signals. Thus, knowing κ and N, the average diameter d of the contacting asperities can be obtained easily. An example of a diffracted signal from the fatigue crack grown under constant ΔK conditions will also be discussed in the following section. In summary, we may thus view the through transmission signal as a first approximation to the scattering problem described by Eq. (6), whereas the diffracted signal should be viewed as a second approximation. For the interested reader, the details of the theory are extensively described in Ref. 18.

53

Crack Tip Shielding by Contacting Asperities

The transmission coefficient as well as the diffracted signals of ultrasonic waves from a crack with partially contacting surfaces grown at constant ΔK have been determined (5). The essential result is that the comparison of theory and experiment provides two quantities which describe the contacting asperities. $\kappa(x)$ is obtained from through transmission experiments; it is the distributed spring constant which contains information on the dimensions and areal density of the contacting asperities. N, the areal density of the contacting asperities, is obtained from diffraction experiments. It is the objective of this brief section to provide information on the effect of these contacting asperities on the stress intensity factor, K_I(local), which shields the crack tip from the externally applied stress intensity factor K_I(global) (4). For this purpose, we make the assumption that each individual asperity which comes in contact with the opposite crack surface produces a stress intensity factor, K_{IS}(local) which is given by (21)

$$K_{IS}(\text{local}) = \frac{2^{1/2}}{(\pi C)^{3/2}} P_S \frac{1}{[1+(z/C)^2]} \tag{7}$$

where C is the nearest distance between the contact and the crack tip, P_S is the load carried by the contact, and z is the coordinate along the crack front with its origin at the closest point to the contact. Superposition of the stress intensity factors of a row of such individual contact points parallel to the crack front and assuming that the distance between two neighboring contact points is also C, yields the stress intensity factor contribution (5)

$$dK_I(\text{local}) \approx (\tfrac{2}{\pi})^{1/2} \frac{P_S}{C^{3/2}} \tag{8}$$

which is identical to

$$dK_I(\text{local}) = (\tfrac{2}{\pi})^{1/2} \frac{\Delta P}{BC^{1/2}}, \tag{9}$$

a result obtained by Beevers et al. (4) for a long, thin strip of contacting material, where $\Delta P/B$ is the load on the strip per unit length for a given sample thickness B. The load ΔP in Eq. (9) on the strip can be related to the static stress across the partially closed crack, σ_o, schematically shown in Fig. 1, as

$$\Delta P = dP = B\sigma_o dx \tag{10}$$

where x is a coordinate perpendicular to the crack front (see Fig. 2). The (average) static stress, σ_o, is related to the acoustic transmission coefficient and thus to κ and N by (14)

$$\sigma_o = (\frac{\kappa}{k^*E})^2 (\frac{\pi}{N}) P_m, \tag{11}$$

where $k^* = 2$, E the elastic modulus, and p_m is the "flow pressure" of the material (usually three times the ultimate tensile strength).

54

Combining Eqs. (9)-(11) and integrating over the closure zone thus yields the total local stress intensity factor, which provides the shielding,

$$K_I(local) = (2\pi)^{1/2} \frac{P_m}{(k*E)^2} \int_0^\infty \frac{\kappa^2(x)dx}{N(x)x^{1/2}} \tag{12}$$

Previously (5), Eq. (12) has been applied to a fatigue crack grown under constant ΔK in Al7075 T651. $\kappa(x)$ in this case was found to be approximately

$$\kappa(x) = \kappa_0 exp(-\beta x) \tag{13}$$

and $N(x) = N$, independent of x, so that

$$\int_0^\infty dK_I(local) = K_I(local) = \pi \left(\frac{\kappa_0}{k*E}\right)^2 \frac{P_m}{N\beta^{1/2}} . \tag{14}$$

Evaluation of the acoustic data yielded in this case $K_I(local) \approx 6.8$ MPa m$^{1/2}$. The compressive static stress at the crack tip was found to be $\sigma_{omax} = 340$ MPa. The average distance, C, between contacting asperities was determined to be C $\approx$ 70 μm and the contact diameter at the crack tip to be $d_{max} \approx 35$ μm. The two quantities σ_0 and d fall off with distance x away from the crack tip as

$$\sigma_0 = \sigma_{omax}exp(-2\beta x) \tag{15}$$

and

$$d = d_{max}exp (-\beta x), \tag{16}$$

respectively.

We believe that the values obtained are reasonable. Particularly the estimated σ_{omax} is of the magnitude expected. For an elastic-perfectly plastic material we would estimate σ_{omax} to be the (compressive) yield stress of the material ($\approx$ 500 MPa). Due to stress relaxation at the crack tip as well as fatigue softening we would expect a value that is somewhat smaller. In addition, C is expected to be of the same magnitude as the grain size, which is as calculated. $K_{Imax}(local)$ is roughly 40% of $K_{Imax}(global)$, the maximum stress intensity factor applied during the growth of the crack.

The Effect of an Overload on Crack Tip Shielding

The large increase in spatial resolution of the acoustic probe prompted us to apply this technique to fatigue cracks in Al2024-T351, which was exposed to an overload. Earlier experiments (22) in which the whole crack was illuminated by the acoustic wave clearly showed that an overload created a high closure stress intensity factor which was responsible for crack retardation after the overload. These measurements (22) also showed that the closure stress intensity factor started to decrease, as the crack continued to grow. With the new technique, we expect to obtain detailed information about the geometry of the overload region and the stresses developed, as well as the changes in crack tip shielding.

55

For the overload application, discussed here, the crack was grown for
0.4×10^5 cycles at a constant $\Delta K_1 \approx 12.2$ MPa $\sqrt{m}$. At that point an overload block of $\Delta K_2 = 24.5$ MPa $\sqrt{m}$ was applied for 21 cycles, after which the
test was continued at ΔK_1. As shown in Fig. 3, the crack was retarded for
about 1.3×10^5 cycles and then started to grow again. Acoustic measurements were performed after a total of about 1.9×10^5 cycles were applied,
which is during the second growth period of the crack. We thus expected
that relatively little crack tip shielding was left from the overload
block.

Figure 4 shows the experimental results of the frequency dependent
transmission coefficient; in Figs. 5 and 6 the frequency dependent diffraction signals at $\Theta = \pm 45°$, respectively, are plotted as a function of
the positioning of the transducers, as indicated in Fig. 2. At small x,
the transducer is probing the unbroken ligament of the compact tension
specimen, as indicated by a transmission coefficient of 1 and the small
diffracted signals. In the closure region the transmission coefficient
decreases slowly and both the diffraction signals show a peak. Such a
peak in the diffraction signal has been observed before for a crack grown
under constant ΔK conditions (5). In addition, the transmission coefficient as well as the diffraction signals show a peak at a position x that
corresponds precisely to the location of the crack where the overload
block was applied.

As a first step to evaluate the data we chose a $\kappa(x)$ such that
Eq. (6) represents the best fit to the transmission data. We chose

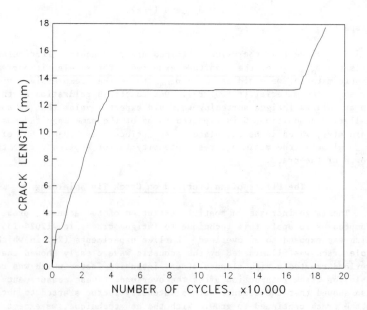

Figure 3 - Crack length versus number of fatigue cycles, indicating
crack growth retardation.

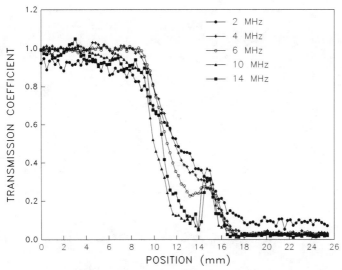

Figure 4 – Ultrasonic transmission through the partially closed fatigue
crack for various frequencies--experimental results. The
peak at a position of about x = 15 mm is caused by the
overload block.

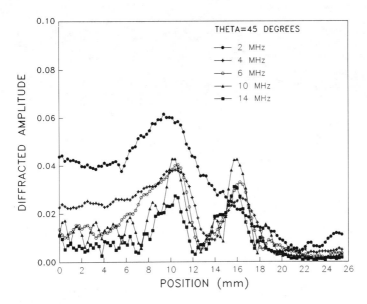

Figure 5 – Signal diffracted at θ = 45° by the partially closed crack
for various frequencies--experimental results.

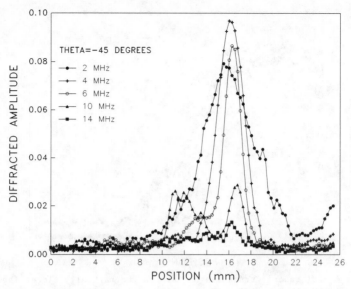

Figure 6 - Signal diffracted at θ = -45° by the partially closed crack and for various frequencies--experimental results.

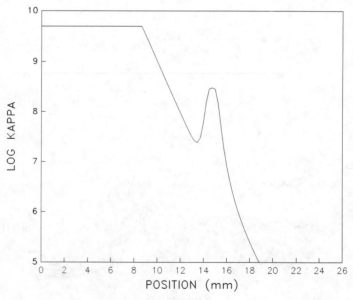

Figure 7 - Selection of κ(x) as a function of x.   κ(x) in MPa m^{-1}.

$$\kappa(x) = \kappa_o e^{-\beta x} + \frac{\kappa_1}{1 + \left[\frac{2(x-\delta)}{\gamma}\right]^4} \qquad (17)$$

where β describes the decay of the spring constant in the closure zone (5). δ is the distance of the crack tip to the position where the over-load was applied, and γ the width of the overload region, taken as the width at half the amplitude of the peak in the transmission coefficient. At $x = \delta$ the amplitude of the spring constant due to the overload is κ_1. Figure 7 shows this function, Eq. (17), for the following parameters

$\kappa_o = 1 \times 10^9$ MPa m^{-1}
$\beta = 1.2$ mm^{-1}
$\kappa_1 = 3 \times 10^8$ MPa m^{-1}
$\alpha = 4.7$ mm
$\gamma = 1$ mm.

The values for κ_o and β are in the range similar to those obtained before for a fatigue crack grown under constant ΔK conditions. α, the crack growth after the local application, was obtained from Fig. 3. γ was estimated from the peak width in Fig. 4 at a position of roughly 15 mm; its value corresponds closely to the calculated plastic zone size in plane strain during the application of the overload. The resulting frequency dependent transmission coefficient is shown in Fig. 8, as calculated from Eq. (6), using the $\kappa(x)$ given in Eq. (17). Comparing Fig. 8 with the experimental results, shown in Fig. 4, indicates fair agreement with some obvious deviations in details. Although the contact due to the overload

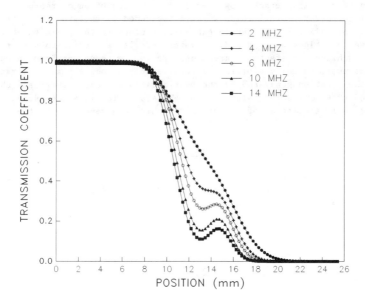

Figure 8 - Ultrasonic transmission through a partially closed crack for various frequencies--theoretical results based on $\kappa(x)$ from Figure 7.

can be clearly seen at the high frequency end of the available spectrum, the amplitude of the calculated peak is smaller than the experimentally observed one. At present we believe that this is caused by having less spatial resolution in the model than we actually have in the experimental set-up. Surprising to us is the result, obtained from Fig. 7, that κ_1 in the overload region is almost of the same size as κ_0 in the vicinity of the crack tip. Evaluating Eq. (11) under the assumption of N being the same in the overload region and at the crack tip one would conclude that the "residual stress" σ_0 at the overload locations is about 1/10 of that at the crack tip which is, using the results obtained earlier (5), about 70% of the yield stress of the material. At the present time, however, we know very little about N(x), which is to be obtained from the diffracted signals (5). The evaluation of N(x) is still in progress. However, we can clearly say that the experimentally observed diffracted signal amplitudes obtained from the overload region are surprisingly large with respect to those obtained from the crack tip region particularly if one compares the signals obtained under $\theta = -45°$ (see Fig. 6). It appears from our previously obtained results (5) on the signal amplitude diffracted at the crack tip, that the present results indicate a relatively low areal contact density N in the overload region. Equation (6), which allows us to calculate the diffracted signal amplitude breaks down for $N < 200$ contacts/cm^2, so that we cannot use it at present to calculate the high amplitudes observed and thus cannot estimate N(x). Future research will concentrate to resolve this problem.

Conclusions

Acoustic transmission and diffracted signals obtained in the closure region of a fatigue crack show clearly the effects of an overload block. In a first order approximation we have succeeded in explaining the transmission signals as being due to extensive contact at the position where the overload block was applied. Several details have still to be worked out to achieve satisfactory agreement. So far, we have not been successful, however, to fully explain the surprisingly strong signals diffracted from the location generated by the overload block. Extensive research is now underway to resolve this problem. This solution is a necessary input in the full calculation of the crack tip shielding generated by the overload block.

Acknowledgement

Ames Laboratory is operated for the U. S. Department of Energy by Iowa State University under contract no. W-7405-ENG-82. This work was supported by the Office of Basic Energy Sciences, Division of Materials Sciences.

References

1. W. Elber, Damage Tolerance in Aircraft Structures, ASTM STP 486 (Philadelphia: Am. Soc. Test. Mat., 1971), 230.

2. C. Q. Bowles and J. Schijve, Fatigue Mechanisms: Advances in Quantitative Measurement of Physical Damage, ASTM STP 811 (Philadelphia: Am. Soc. Test. Mat., 1983), 400.

3. S. Suresh and R. O. Ritchie, Scripta Met., 17, (1983) 595.

4. C. J. Beevers, et al., Engr. Fract. Mechanics, 19 (1984), 93.

5. O. Buck, D. K. Rehbein, and R. B. Thompson, "Crack Tip Shielding by Asperity Contact as Determined by Acoustic Measurements," submitted to Engr. Fract. Mechanics.

6. S. Golan, et al., J. Nondestructive Evaluation, 1 (1980), 11.

7. O. Buck and B. R. Tittmann, Advances in Crack Length Measurement, ed. C. J. Beevers (Cradley Heath, Warley, West Midlands, U. K. Engineering Materials Advisory Services, 1982), 413.

8. S. Golan and R. Arone, New Procedures in Nondestructive Testing, ed. P. Höller (Berlin, Heidelberg, New York, Springer Verlag, 1983), 587.

9. R. B. Thompson, C. J. Fiedler and O. Buck, Nondestructive Methods for Materials Property Determination, eds. C. O. Ruud and R. E. Green (New York: Plenum Press, 1984), 161.

10. R. B. Thompson and C. J. Fiedler, Review of Progress in Quantitative Nondestructive Evaluation, Vol. 3A, eds. D. O. Thompson and D.E. Chimenti (New York: Plenum Press 1984), 207.

11. O. Buck and R. B. Thompson, Fatigue 84, vol. II, ed. C. J. Beevers, (Cradley Heath, Warley, West Midlands, U.K.: Engineering Materials Advisory Services), 667.

12. J.-M. Baik and R. B. Thompson, J. Nondestructive Evaluation, 4 (1984), 177.

13. Y. C. Angel and J. D. Achenbach, J. Appl. Mech., 107 (1985), 33.

14. O. Buck, R. B. Thompson and D. K. Rehbein, J. Nondestructive Evaluation, 4 (1984), 203.

15. D. K. Rehbein, R. B. Thompson and O. Buck, Review of Progress in Quantitative Nondestructive Evaluation, Vol. 4A, eds. D. E. Thompson and D. E. Chimenti (New York: Plenum Press, 1985), 61.

16. R. B. Thompson, O. Buck and D. K. Rehbein, "Elastic Wave Interactions with Partially Contacting Interfaces: Status of Theory and Application to Fatigue Crack Closure Characterization", Proceedings on Solid Mechanics Research for QNDE, eds. J. D. Achenbach and Y. Rajapakse (Dortrecht: Martinus Nijhoff Publishers) in press.

17. O. Buck, D. K. Rehbein, and R. B. Thompson, "Determining Crack Tip Shielding by Means of Acoustic Transmission and Diffraction Measurements", Review of Progress in Quantitative Nondestructive Evaluation, Vol. 6, eds. D. O. Thompson and D. E. Chimenti (New York: Plenum Press), in press.

18. R. B. Thompson, O. Buck and D. K. Rehbein, "The Influence of Asperity Contact on the Scattering of Elastic Waves from Fatigue Cracks", _Proceedings 10th National Congress of Applied Mechanics_, (Austin, TX) in press.

19. O. Buck, R. B. Thompson and D. K. Rehbein, "Using Acoustic Waves for the Characterization of Closed Fatigue Cracks", _Proceedings of the International Symposium on Fatigue Crack Closure_, ASTM STP (Philadelphia: Am. Soc. Test. Mat.) in press.

20. B. A. Auld, _Wave Motion_ 1 (1979), 3.

21. H. Tada, P. C. Paris and G. R. Irwin, _The Stress Analysis of Cracks Handbook_, (St. Louis: Del Research Corporation, 1973).

22. O. Buck, J. D. Frandsen, and H. L. Marcus, _Fatigue Crack Growth under Spectrum Loads_, ASTM STP 595 (Philadelphia: Am. Soc. Test. Mat., 1976), 101.

FATIGUE CRACK GROWTH WITH SINGLE OVERLOAD:

MEASUREMENT AND MODELING

D. L. Davidson, S. J. Hudak, Jr., and R. J. Dexter

Southwest Research Institute
6220 Culebra Road
San Antonio, TX   78284

ABSTRACT

This paper compares experiments with an analytical model of fatigue crack growth under variable amplitude. The stereoimaging technique was used to measure displacements near the tips of fatigue cracks undergoing simple variations in load amplitude - single overloads and overload/underload combinations. Measured displacements were used to compute strains and stresses were determined from the strains. Local values of crack driving force (ΔK effective) were determined using both locally measured opening loads and crack tip opening displacements. Experimental results were compared with simulations made for the same load variation conditions using Newman's FAST-2 model. Residual stresses caused by overloads, crack opening loads and growth retardation periods were compared.

Acknowledgement

Support for this research came from NASA, Langley Research Center, under Contract NAS1-17641. We are grateful to Jim Newman for many helpful discussions.

INTRODUCTION

Fatigue crack growth is often studied under constant amplitude loading conditions, both because of the ease of experimentation and because of the complexity of understanding the phenomena associated with fatigue. However, in actual use, structures frequently undergo a loading amplitude which is variable. Fatigue crack growth rates under variable amplitude loading are no longer directly related to the level of stress intensity factor; therefore, crack velocities are not predictable from constant amplitude laboratory experiments. There have been several attempts to handle the effects of variable amplitude loading on crack growth rates through the use of relatively simple modeling procedures. The models of Wheeler (1) and Willenborg (2) are the most used because they do compensate fatigue crack growth rates, at least partially, for variations in loading amplitude. A more recent modeling procedure is that of Newman (3), which, although providing a more physically realistic treatment of load interaction effects is also more complex.

The purpose of this paper is to report on a limited number of experiments which obtained detailed information on the deformation attending fatigue crack growth using highly simplified loading variations, which were intended to simulate the largest effects of complex spectral loading. Single overloads of several amplitudes were studied, and in one case, an overload followed by an underload. Crack growth rate was measured throughout the period affected by the variable amplitude loading. Displacements within the plastic zone of the crack were measured at each step in the load variation, and near the minimum crack growth rate which resulted from application of the variable load. The load necessary to open the crack, often termed "closure", was also measured local to the crack tip throughout the course of these experiments.

Experimental results were compared to predictions from Newman's analysis of variable amplitude loading, which has been reduced to practice in the FAST-2 computer program (4). Only one of several cases studied is shown in detail in the present paper, while more extensive results have previously been reported in Refs. 5 and 6.

EXPERIMENTAL METHODS

The terms used to describe the magnitudes of the loading variations are noted in Fig. 1, and the actual cases studied are listed in Table I. The procedures used to obtain information during the loading histories is shown schematically in Figs. 2 and 3. All experiments were performed in a cyclic loading machine designed to fit in the specimen chamber of a scanning electron microscope (SEM) (7). The high resolution and large depth of field of the SEM can be obtained when this cyclic loading stage is used. Photographs of the crack tip region were made at magnifications of between 400 and 1000 times. Under constant amplitude loading, just prior to the application of the variable loading sequence, the crack opening load was measured. The stereoimaging technique (8) was used to determine strains within the plastic zone of the crack, as well as crack opening displacements behind the crack tip, by comparing photographs made at #1 and #2 of Fig. 2. Plastic zone of the constant amplitude crack is shown in Fig. 3.

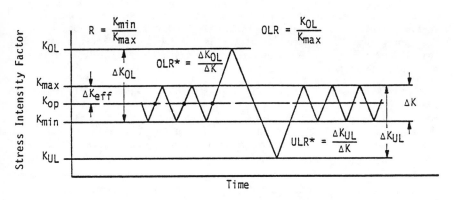

Figure 1. Definitions of terms describing fatigue crack growth and load range variation severity.

Table I. Conditions Tested

Experiment	R	ΔK, MPa$\sqrt{m}$	OLR*	ULR*
1	0.16	6.9	2.15	1.0
2	0.22	6.0	2.85	1.0
3	0.50	7.2	3.0	2.0

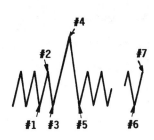

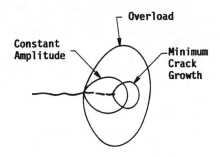

Figure 2. Load sequence showing points at which photographs were made for analysis.

Figure 3. Schematic of plastic zone size, shape and location at the points of analysis.

The load required to open the crack to the tip in Mode I was determined by stereoimaging a photograph made at minimum load with those made at various levels of load between minimum and maximum, with the load axis of the photographs in the same direction as the eye axis. In this way, the location of the open crack relative to the crack tip was directly measured from the photograph.

Application of the spike overload, #4 in Fig. 2, caused the crack to remain open when the load was again reduced to the minimum, #5, and it resulted in a plastic zone larger than that caused by constant amplitude loading, as shown in Fig. 3. Measurements of displacements were made on the loading part of the cycle by comparing photographs made at #3 and #4, and on the unloading part of the cycle, by comparing photographs made at #4 and #5.

In one of the experiments, an underload cycle was applied after the overload cycle. This was done from R = (K_{min}/K_{max}) = 0.5 by reducing the load to zero. Two additional analyses were required for this case, one which measured displacements on the unloading part of the underload cycle, and another for the reloading part of the cycle.

After application of the variable amplitude cycles, the load levels were returned to those used for constant amplitude crack growth and cycling was continued. For the single spike overloads, this resulted in a short burst of rapid crack growth, followed by a period of slower crack growth, often termed the crack retardation period. Crack opening loads were determined periodically as the crack advanced into the overload zone, and near the minimum in crack growth, analyses of the crack tip were once again made by comparing photographs taken at minimum and maximum load, #6 and #7 of Fig. 2. A smaller plastic zone size resulted for the crack in this location, as is illustrated in Fig. 3.

EXPERIMENTAL RESULTS

Opening of the crack with increasing load was found to be highly nonlinear, as is shown in Fig. 4. Furthermore, the level of the crack opening load was found to be dependent upon both the magnitude of cyclic stress intensity factor, ΔK, and the R ratio. Although only a small amount of data are shown in the figure, an extensive data base substantiates these observations (6). These data clearly show the importance of local measurements of this parameter. Similar data were generated for crack opening measurements made as the crack traversed through the overload growth retardation zone.

Strain distributions near the crack tip before, during, and after the overload cycle are illustrated in Fig. 5. Strain was always found to peak at the crack tip, decreasing with distance into the plastic zone. Comparison of the normalized strain magnitudes ahead of the crack tip before and after the overload is shown in Fig. 6. Normalized strain is the measured value at a given distance/crack tip strain. Since these strain distributions are essentially the same, the cracks are considered to be demonstrating similitude. Linear elastic fracture mechanics is predicated on using K to describe cracks having self-similar stress fields. Therefore, it should

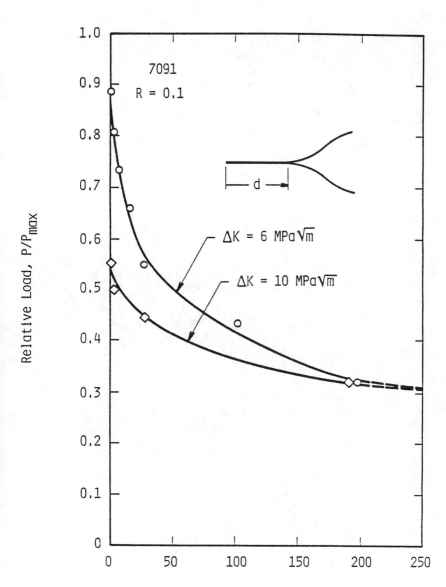

Figure 4. Length of crack remaining closed versus relative cyclic load for different applied ΔK values illustrating the manner in which the crack "peels" open.

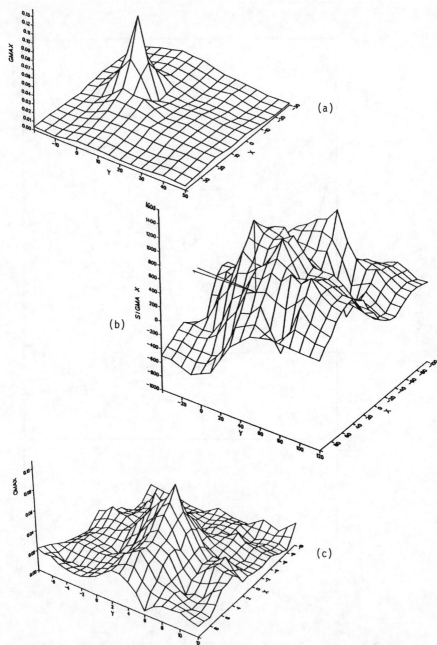

Figure 5. Distribution of maximum shear strain (a) before, (b) at the peak of the overload, and (c) during subsequent crack growth after the overload. X and Y are in micrometers. Crack tip is beneath the value of maximum strain in the orientation schematically shown in (b).

68

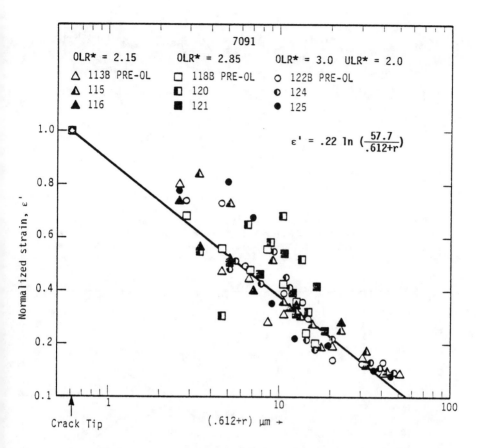

Figure 6. Comparison of the effective strain distributions before and after the overload cycle. Similitude appears to be approximately preserved in spite of the presence of the overload residual stress field.

be possible to describe crack growth both before and after the overload using K.

The plastic zones formed by the crack before and after the application of the load variation are compared in Fig. 7. The overload/underload combination results in less of a contraction of plastic zone size than do overloads without underloads. The distance ahead of the crack tip is to scale, but not distances in the loading direction.

For crack growth under constant amplitude loading, it has been shown that the rate of crack growth is related to the magnitude of strain at the crack tip (9). As the correlation in Fig. 8 shows, this is also true for crack growth after a load variation. This is another manifestation of the

69

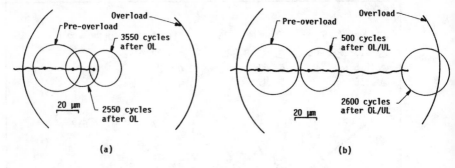

Figure 7. Relative sizes of plastic zones at constant amplitude on the pre-overload cycle, on the overload cycle and at several points along the growth path after the overload. (a) OLR* = 2.15, and (b) OLR* = 3.0, ULR* = 2.0. Plastic zones are shown as near-circular, which is a good average representation.

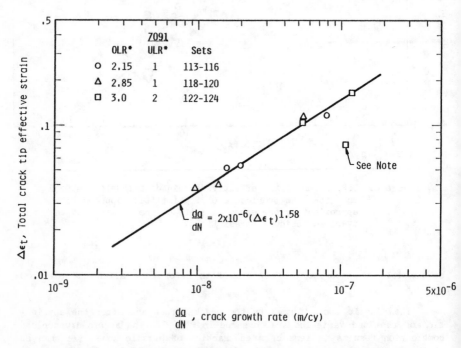

$$\frac{da}{dN} = 2 \times 10^{-6} (\Delta \epsilon_t)^{1.58}$$

Figure 8. Correlation of crack tip strain with crack growth rate. Note: Crack growth rate averaged over 100 μm. Growth rate at the time of measurement of crack tip strain was probably less.

70

similitude being maintained during crack growth through the zone of growth retardation.

Measured crack opening displacements (COD) may also be used to compute a crack driving force, which may be compared to that determined from measured crack opening loads. When COD is plotted against square root of distance behind the crack tip, a straight line may be fit through the data, including the origin, as shown in Fig. 9. The crack tip opening displacement (CTOD) is defined as being the COD at 1 μm behind the crack tip. The experimental driving force, ΔKe, was computed using this CTOD in the equation (10)

$$\Delta Ke = (CTOD\ \Delta\sigma\ E)^{1/2} \qquad [1]$$

where $\Delta\sigma$ = the cyclic stress at the crack tip, as calculated from using the measured crack tip strain and the stable cyclic stress-strain curve. The values of ΔKe thus calculated are shown in Table II, together with the levels of ΔK_{eff} determined from Mode I crack opening loads measured locally, as shown in Fig. 4. Values of ΔKe decrease after the overload cycle, but not as much as the values of ΔK_{eff} determined from the crack opening load; ΔK_{eff} values are consistently smaller than ΔKe values.

Stress changes in the coordinate axes were computed from the changes in strains determined by stereoimaging using the method described in Ref. 11. Changes in the stresses for each step of the load variation cycle were computed at the same position relative the the crack tip and added together to determine the residual stress resulting from that cycle. To reference these changes in computed stresses to zero, it was necessary to divide the stress range determined for the cycle before the overload into tensile and compressive components, and that was done using the measured crack opening loads. These partitioned stresses were then added to those of the cycle of load variation to give a residual stress referenced to zero. One of the residual stress distributions so computed is shown in Fig. 10. As may be seen, there is a zone of residual compressive stress ahead of the crack tip; the size of this region coincides approximately with the plastic zone caused by the overload cycle. For the experiment which included an underload following the overload, the size of the residual compressive region was reduced relative to those experiments having only an overload.

MODELING

The FAST-2 fatigue crack growth simulation model was used to duplicate as closely as possible the experimental results described above. This model uses a strip of tensile ligaments along the crack line which stretch as the crack passes. Displacements are computed using the Dugdale (crack) model which also allows the plastic zone size and crack surface displacements to be obtained. The essential features of this model are shown in Fig. 11, which was taken from Newman's paper (3). The situation at maximum applied stress, Fig. 11(a), shows stretched ligaments within the plastic zone (p) ahead of the crack tip, and along the crack flank (c). The model is for a rigid-perfectly plastic material, so the stress within the plastic zone is constant, as shown in the lower part of the figure. The magnitude

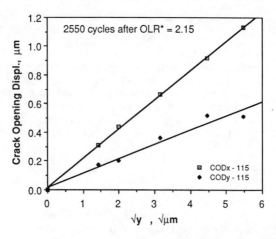

Figure 9. Crack opening displacement in Mode I (squares) and Mode II (diamonds) vs square root of distance behind the crack tip showing linearity of this method of graphing COD.

Table II. Comparison of Measured and Applied Values of ΔK

Analysis	CTOD (μm)	P_{op}/P_{max}	ΔKe MPa√m	ΔK_{eff} MPa√m
1. OLR* = 2.15 ΔK = 6.9 MPa√m R = 0.16				
Before OL	0.32	0.57	5.33	3.52
+2550 cy	0.21	0.52	4.17	3.93
+3550 cy	0.16	0.74	3.62	2.14
2. OLR* = 2.85 ΔK = 6.0 MPa√m R = 0.22				
Before OL	0.20	0.66	4.22	2.58
+1200 cy	0.19	0.90	3.87	0.80
+3500 cy	0.19	0.84	3.89	1.20
3. OLR* = 3.0 ULR* = 2.0 ΔK = 7.2 MPa√m R = 0.5				
Before OL	0.58	0.66	7.2	4.90
+500 cy	0.29	0.65	4.8	5.04
+2600 cy	0.25	0.76	4.7	3.38

ΔKe - derived from CTOD and crack tip strain, Eq. [1].
ΔK_{eff} - derived from opening load measurement.

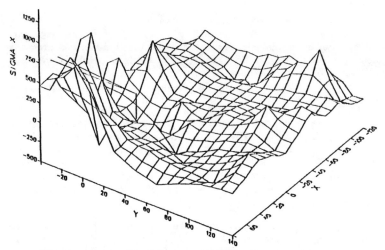

Figure 10. Residual stress field in the direction of loading caused by OLR* = 3.0.

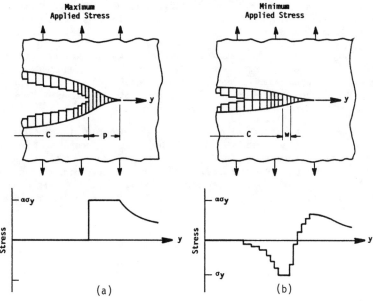

(a)   (b)

Figure 11. Basic concepts in FAST-2 model. (a) A strip of elements within the plastic zone p ahead of a crack of length C loaded in tension. The Dugdale model is used to cancel the stress singularity at the crack tip and give crack opening displacements. The stress distribution shown is that for an elastic-plastic material. As elements break at the crack tip, they remain elongated, so that for minimum stress (b) these elements contact giving rise to closure and the stress distribution shown.

is that of a stress equal to the average of the yield and ultimate stress, σ. When load is decreased towards minimum the stretched ligaments in the wake of the crack come into contact, so that the crack closes prior to reaching minimum load. This causes yielding in compression to shrink to width w, and the stress pattern to have a region of minimum residual stress as shown in the lower part of the figure, which is greatest at, and near, the crack tip.

Newman's model also considers the effect of stress state on the deformation of uncracked elements (within the crack-tip plastic zone) through the use of a constraint factor, α, which can vary from 1 (plane stress) to 3 (plane strain). This factor elevates the materials' flow stress (in tension only) for crack-tip elements, thereby simulating the effect of three-dimensional constraint exerted by elastic material surrounding the crack tip.

The modeling procedure described above has been incorporated in a computer program known as FAST-2, which calculates crack closure stress intensity factors for the loading conditions imposed. The effective stress intensity factor for the crack is then computed from the applied and closure values, and the crack growth rate is determined by entering a measured curve of crack growth rate vs effective stress intensity factor. Using this procedure allows simulation of crack growth to be made under variable amplitude loading.

FAST-2 was used to simulate the experiments performed in order to compare the model to experiment. Only one of the experimental cases, an OLR* = 2.15, is reported here, but the results are similar for all the cases considered. The crack growth rates which occurred before, during, and after the load variation are shown in Fig. 12.

The stress distribution caused by the overload is compared in Fig. 13, which shows a cross section of the stresses computed from measured strains, which were shown in Fig. 10, with the stress distribution of the FAST-2 simulation superimposed. The region of compressive residual stresses determined from experiment is about 4 times larger than that determined using the FAST-2 simulation. This difference is consistent with the differences between measured and predicted crack growth rates in Fig. 12.

Simulation of the crack opening under increasing load is shown in Fig. 14, and is compared to the experimentally determined values. The model predicts a value which is independent of the applied stress intensity factor, while the experimental values are dependent on the level of stress intensity factor. This trend was also found at higher R ratios, and it was not possible to duplicate these results by adjusting the several constants within the simulation.

DISCUSSION

The reason it has been so difficult to predict fatigue crack growth under conditions of variable amplitude loading is because the rate at which a fatigue crack advances is determined by several factors. The level of applied stress intensity factor is probably the most important factor, but

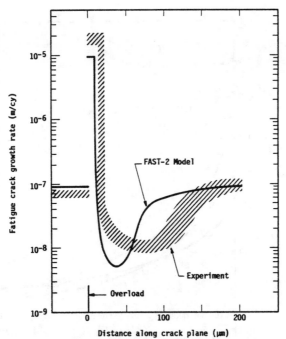

Figure 12. Comparison of experiment with model results, for single overload OLR* = 2.15, R = 0.1. Constraint factor in the model α = 1.9. Results were relatively insensitive to for the range 1 < α < 1.9.

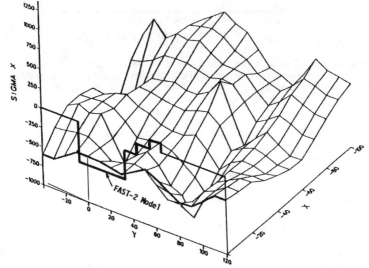

Figure 13. Cross section along the crack path line of the residual stress distribution caused by an OLR* = 2.15 from ΔK = 6.9 MPa$\sqrt{m}$, R = 0.1. Also shown is the stress distribution predicted by FAST-2.

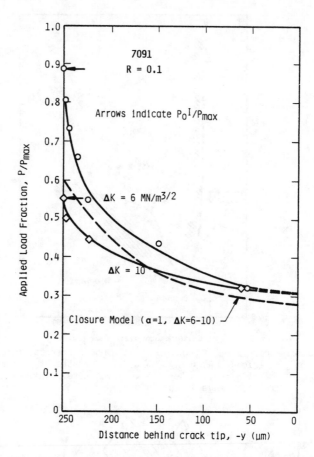

Figure 14. Comparison between the peeling opening of the crack as measured and that predicted by FAST-2. Experimentally, opening load fraction is dependent on ΔK, whereas FAST-2 predicts a constant value.

this applied stress intensity is modified by the extent of crack closure experienced by the crack. Plasticity induced crack closure is caused, it is thought, by two factors: the irreversible stretching of ligaments along the crack wake, and the region of compressive residual stress ahead of the crack tip. Under constant amplitude loading, the level of closure is established by the presence of both factors. It might be possible, at least in principal, to attribute some portion of the closure load to each factor, but a method for doing this has not yet been found. However, upon application of a variable load, the closure condition which had been established under constant amplitude is disrupted, so that the level of closure caused by each factor is changed. As has been shown (12), an overload usually causes the crack to remain permanently open, thereby nullifying any effect of the previous wake at that point. The zone of residual stress ahead of the crack tip is increased in size by an overload, Figs. 7 and 10, although the level apparently remains at approximately the yield stress. Therefore, just after application of the overload, there is no closure and the zone of residual compressive stress ahead of the crack tip is disproportionately large.

Immediately following an overload, the rate of crack growth increases, reflecting the loss of the previous crack wake, but as the crack lengthens into the overload zone, and a new wake is established, the rate of crack growth decreases to a level below that which occurred before the overload, which either reflects the increased stretching of the ligaments in the wake of the crack caused by the overload, or the increased residual stress in the zone through which the crack is growing, or both. Whatever the resultant crack wake and residual stress effects, similitude, as defined in Fig. 6, is apparently re-established within a short distance of the overload. The fact that there is similitude probably means that some balance between the wake and residual stress conditions similar to a crack growing under constant amplitude conditions has been established.

Effective driving force values experimentally derived after the overload and overload/underload were consistently smaller when the load to open the cracks was used as compared to use of the crack opening displacement and crack tip strain (Eq. 1). One reason for this may be that use of CTOD is more of an averaging technique than is measurement of the crack opening load. Thus, CTOD may give more of an average through-the-thickness value than does opening load.

Another change caused during crack growth through the overload region is that of the change in effective R ratio. Because of the zone of compressive stresses ahead of the overload cycle, subsequent growth must be occurring under stress ratio conditions different than before the overload. A shift to lower R would decrease the rate of fatigue crack growth in addition to that caused by the decrease in ΔK_{eff}.

The FAST-2 model simulates the effects of both residual stretch in the crack wake and the residual stress ahead of the crack tip, and it does so in a computationally efficient manner. However, results using the model do not correspond with experimental findings very well, particularly the effect of K magnitude on the level of crack opening load.

SUMMARY AND CONCLUSIONS

1. Crack opening load is very sensitive to the method by which it is measured, particularly in the near-threshold region, and varies with the magnitude of ΔK.

2. An overload cycle causes a zone of compressive residual stress of about yield stress magnitude to be developed ahead of the crack tip; the size of this zone exceeds that of the measured cyclic plastic zone and is dependent on the magnitude of the overload.

3. A fatigue crack growing through an overload zone quickly re-establishes similar relations between crack tip strain, strain distribution ahead of the crack tip, and crack opening displacement as were found prior to the overload.

4. The driving force, ΔK_{eff} experienced by the crack is decreased after an overload. The magnitude of ΔK_{eff} decreases as the severity of the overload increases. This decrease in ΔK_{eff} was derived both from the level of the measured crack opening load, and from an analysis of the measured crack opening displacement and crack tip strain. An underload following the overload tends to nullify the effect of the overload.

5. A strip yield model of the effect of variable amplitude loading, as implemented in the FAST-2 program, captures the essence of the experimental findings, but not the details. Specifically, the spacial extent of crack growth retardation was underestimated, the crack opening loads predicted were not ΔK dependent, and the zone of residual stress was smaller than that predicted.

REFERENCES

1. O. E. Wheeler, "Spectrum Loading and Crack Growth," Journal of Basic Engineering, 94 (1972), 181-186.

2. J. Willenborg, R. M. Engle and R. A. Wood, "A Crack Growth Retardation Model Using an Effective Stress Concept" (Air Force Flight Dynamics Lab. Report AFFDL-TM-71-1-FBR, 1971).

3. J. C. Newman, Jr., "A Crack-Closure Model for Predicting Fatigue Crack Growth Under Aircraft Spectrum Loading" Methods and Models for Predicting Fatigue Crack Growth Under Random Loading, (Philadelphia, PA, ASTM STP 748, 1981), 53-84.

4. J. C. Newman, Jr., FAST-2 Computer Program. Available from COSMIC, University of Georgia, Athens, GA 30602.

5. D. L. Davidson and S. J. Hudak, Jr., "Alterations in Crack-Tip Deformation During Variable Amplitude Fatigue Crack Growth" (Philadelphia, PA, ASTM, 1985, in press).

6. S. J. Hudak, Jr. and D. L. .Davidson, "The Dependence of Crack Closure on Fatigue Loading Variables" (Philadelphia, PA, ASTM, 1986, in press).

7. D. L. Davidson and A. Nagy, "A Low Frequency Cyclic Loading Stage for the SEM," Journal of Physics E, 11 (1978), 207-210.

8. D. R. Williams, D. L. Davidson, and J. Lankford, "Fatigue-Crack-Tip Plastic Strains by the Stereoimaging Technique," Experimental Mechanics, 20 (1980), 134-139.

9. D. L. Davidson and J. Lankford, "Fatigue Crack Tip Mechanics of a Powder Metallurgy Aluminum Alloy in Vacuum and Humid Air," Fatigue of Engineering Materials and Structures, 7(1) (1984), 29-39.

10. K. S. Chan, "Local Crack-Tip Field Parameters for Large and Small Cracks: Theory and Experiment," <u>Small Fatigue Cracks</u>, ed. R. O. Ritchie and J. Lankford (Warrenville, PA, TMS-AIME, 1986), 407-425.

11. D. L. Davidson, D. R. Williams, and J. E. Buckingham, "Crack-Tip Stresses as Computed from Strains Determined by Stereoimaging," <u>Experimental Mechanics</u>, 23(2) (1983), 242-248.

12. J. Lankford and D. L. Davidson, "The Effect of Overloads Upon Fatigue Crack Tip Opening Displacement and Crack Tip Opening/Closing Loads in Aluminum Alloys," <u>Advances in Fracture Research</u>, ed. D. F. Francois (Pergammon Press, Oxford, 1980), 899-906.

Mechanisms of Elevated Temperature Fatigue Crack Growth

in Inconel 718 as a Function of Stress Ratio

S. Venkataraman[*] and T. Nicholas[**]

[*] Materials Engineer, Systran Corporation - AFWAL/MLLN, Wright-Patterson Air Force Base, Ohio 45433
[**] Senior Scientist, AFWAL/MLLN, Wright-Patterson Air Force Base, Ohio 45433

Abstract

Constant load amplitude as well as major / minor cycle tests were conducted on Inconel 718 at 649°C using center cracked tension specimens, M(T). Detailed fractographic investigation was conducted on selected specimens using a scanning electron microscope to study the associated micromechanisms. A strong relationship was observed between load ratio, mean stress level and crack growth micromechanisms which was used to interpret the major / minor cycle fatigue crack growth behavior. A transition from cycle dependent to time and environment dependent behavior was observed as R was increased up to a value of 0.8 to 0.9.

Introduction

Modern Air Force gas turbine engine design and life management procedures such as Retirement-for-cause (RFC) [1] and Engine Structural Integrity Program (ENSIP) [2] require life prediction analysis based on crack growth rates calculated from pre-existing flaws using fracture mechanics concepts. In the past, such predictions of crack growth have been generally based on modeling and understanding derived from crack growth studies associated with cyclic and sustained loading under temperatures and environmental conditions representative of engine operations [3]. Effects of variations in frequency, load ratio (R) and sustained load hold time on crack growth rates have also been studied. With the addition of interaction effects, the results of these fundamental studies have been incorporated into models to address crack growth in specific engine spectrum and provide more meaningful life prediction methods for engine components [4].

Several investigators [5-7] have shown that the elevated temperature fatigue crack growth rate of nickel base superalloys decreases with increasing frequency in the regime in which environmental and time dependent crack growth processes are predominant. Fractography in these studies revealed that this decrease in fatigue crack growth rate is accompanied by a change in fracture mode from intergranular to transgranular. In addition, Weerasooriya [6] and Solomon and Coffin [8] have shown that for test frequencies above a certain level the fatigue crack growth rate is independent of frequency. In the intermediate frequency regime, several crack growth mechanisms operate simultaneously leading to interaction effects and transient phenomena in crack growth behavior. This region can be important in typical load spectra for gas turbine engine disks when evaluating crack growth behavior [9].

Studies of crack growth at elevated temperatures where time dependent processes, such as creep and environmental interactions,

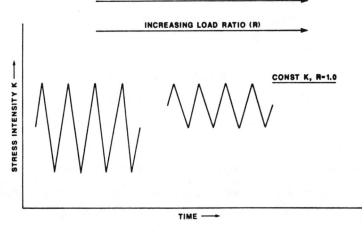

Figure 1. Schematic illustrating relationship between load ratio and crack growth mechanisms.

become more active show generally that crack growth rate (da/dN) for a given stress intensity range, ΔK, increases with increase in R [10-14]. However, the above mentioned frequency effect on crack growth behavior may alter this trend. This is rationalized by two different mechanisms operable at the two extremes of R as shown schematically in Fig. 1. At low R, the crack growth behavior is essentially cycle dependent and mean stress has little influence on growth rate. At high R values, however, mean stress becomes much more important as the mode of crack growth gradually changes towards purely time dependent, typical of that observed in sustained load crack growth. At high R values, therefore, there are two contributions to the crack growth process, namely, the cyclic contribution due to stress amplitude and the time dependent contribution due to mean load.

The elevated temperature fatigue crack growth behavior of Inconel 718 has been studied in this investigation as a function of load ratio. This paper presents various crack growth micromechanisms observed as a function of stress ratio and their relationship to measured crack growth rates. Further, this understanding is extended to interpret the major / minor cycle fatigue crack growth in Inconel 718.

Experimental Procedure

A series of constant load amplitude fatigue crack growth tests was conducted on center cracked tension, M(T), specimens of Inconel 718 nickel base superalloy. The alloy 718 was heat treated following the procedure given in Table I. Test specimens of dimensions shown in Fig. 2 were machined out of the heat treated material. The center hole and the starter notch were machined using an electric discharge milling technique.

Stress ratio values for these tests were chosen between 0.1 and 0.9 and were maintained constant during individual tests. Frequencies from 1000 to 0.001 Hz were studied. All tests were conducted at a temperture of 649°C (1200°F). The test specimen was maintained at this temperature using a resistance heated oven equipped with quartz viewing ports through which optical crack length could be periodically monitored with the aid of a travelling microscope. Primary crack length data were obtained using an a-c electric potential technique [15]. Derived crack lengths were verified through optical surface crack length measurements.

Fatigue crack growth tests at 100 Hz and above were conducted using an electro-pneumatic shaker system. A pre-load was applied on the test sample and the cycling was done using the shaker. Tests at 10 Hz and below were performed using an MTS servo-hydraulic test machine with a computer controlling the test parameters and acquiring and reducing the data.

The major / minor cycle fatigue tests referred to in the latter section of this paper were conducted using the waveform shown in Fig. 3. It consisted of a 1 Hz low frequency major cycle loading with a 180

Table I

Heat Treatment for Inconel 718

Step 1. Solution at 968°C for 1 hour, then air cool to 718°C.

Step 2. Age harden at 718°C for 8 hours, then furnace cool at 56°C/hour to 621°C.

Step 3. Age harden at 621°C for a total aging time in step 2 and step3 of 18 hours.

Step 4. Air cool to room temperature.

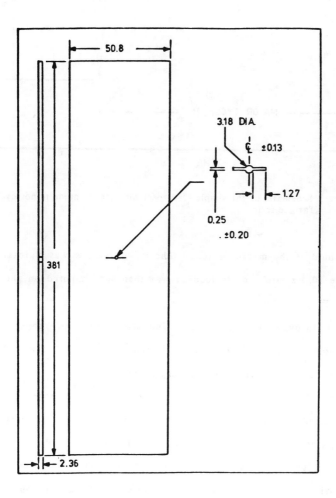

Figure 2. Schematic of center cracked tension specimen, M(T). Dimensions are in millimeters.

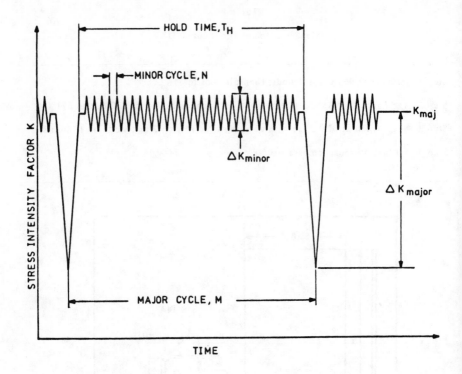

Figure 3. Schematic of the simple load spectrum used in major/minor cycle
fatigue tests.

seconds hold at the maximum load. The major cycle R was maintained at
0.1. The 10 Hz minor cycle loading was then superimposed on the major
cycle hold time load. Stress intensity was used as the controlling
parameter throughout these tests. The stress intensity range for the
major cycle, ΔK_{major}, was maintained constant during individual tests.
This was done by decreasing the major cycle load as the crack length
increased. Likewise, the minor cycle stress intensity range, ΔK_{minor},
was initiated at a higher level and was decreased linearly with
increasing crack length. Further details are provided elsewhere [16].

To evaluate the crack growth micromechanisms as a function of
stress ratio, detailed fractographic analyses were carried out on all
fractured test specimens using a scanning electron microscope.

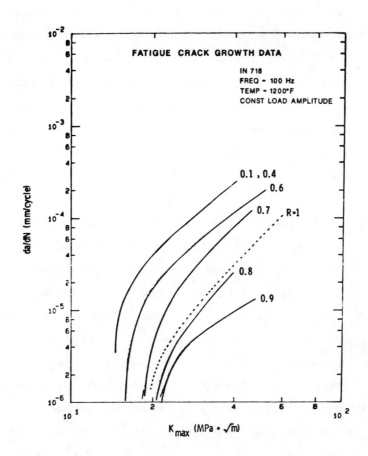

Figure 4. Fatigue crack growth rate (da/dN) versus maximum stress intensity (K_{max}) plots from constant load amplitude tests run under
load ratios (R) between 0.1 and 0.9 and a frequency of 100 Hz.
Sustained load crack growth data are also shown in dotted line
(R = 1.0).

Results and Discussions

The constant load amplitude fatigue crack growth data for Inconel
718 obtained at 100 Hz frequency are shown in Fig 4 in the form of the
crack growth rate (da/dN) versus maximum stress intensity (K_{max}). Under
constant load amplitude, the crack tip stress intensity increases with
increase in crack length. As a result the crack growth rates are

87

obtained over a range of maximum stress intensities from each test. The data for R values of 0.1 and 0.4 overlap while the crack growth rate decreases with further increase in R for any given maximum stress intensity. With increase in R under this condition, the load amplitude decreases resulting in a lower cyclic driving force for the crack growth. The mean stress, on the other hand, increases with increase in R. The sustained load crack growth data (R = 1.0) for Inconel 718 is also given in Fig. 4 in dotted line and the data lie between those for R values of 0.7 and 0.8.

Fractographic examination of these test specimens reveals two types of fracture morphology. At a maximum stress intensity value of 30 MPa$\sqrt{m}$ and 100 Hz frequency, transgranular fracture morphology is observed at R = 0.1 and 0.6 as shown in Fig. 5a through 5d. Striation-like features are also noted in these cases. At a high R value of 0.9, however, totally intergranular fracture mode is observed as indicated in Fig. 6a and 6b. Considerable secondary cracking is also observed in this case.

The constant load amplitude fatigue crack growth data obtained at 10 Hz frequency are given in Fig. 7. Again, the crack growth rate at any given maximum stress intensity studied decreases with increase in R, although less rapidly when compared to that at 100 Hz test frequency. The fracture morphology for the specimen tested at an R value of 0.1 and maximum stress intensity value of 50 MPa$\sqrt{m}$ is of transgranular mode as shown in Fig. 8a and 8b. Fatigue striations are clearly observed indicating a cycle dependent mechanism. At an R value of 0.8 and maximum stress intensity of 40 MPa$\sqrt{m}$, however, the fracture morphology is intergranular as indicated in Fig. 9a and 9b.

This observation along with the increased crack growth under purely sustained load conditions shown in Fig. 4 indicates a transition from cycle dependent to time dependent mechanism occurring in a region corresponding to an R value of 0.8.

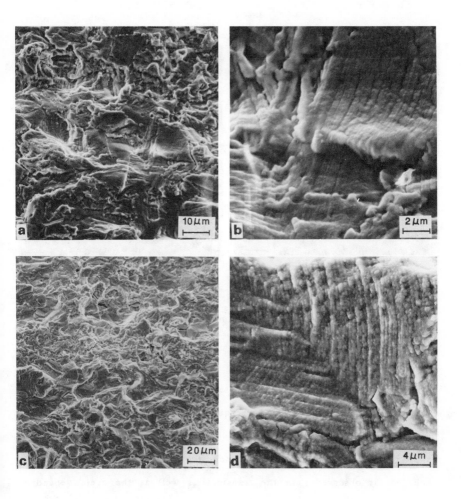

Figure 5. Fracture morphologies observed in Inconel 718 specimen tested at a frequency of 100 Hz and a load ratio of 0.1 (a,b) and 0.6 (c,d). Striation-like features were observed. Crack growth from left to right.

The influence of R on the fatigue crack growth behavior is more easily seen from a crack growth rate (da/dN) versus frequency plot as shown in Fig 10. These curves are plotted for a maximum stress intensity of 30 MPa$\sqrt{m}$. The sustained load crack growth data are shown as to R = 1.0. At low frequencies, the time dependent as well as the environment assisted crack growth predominates. Here, the crack growth

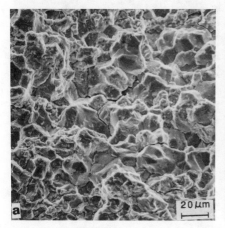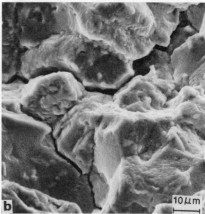

Figure 6. Fracture morphologies observed in Inconel 718 specimen tested at a frequency of 100 Hz and a load ratio of 0.9. Totally intergranular fracture mode was observed. Crack growth from left to right.

increases with increase in R as observed by other investigators [10-14]. On the other hand, at high frequencies, a totally cycle dependent mechanism operates as seen for R = 0.1 for frequencies in excess of approximately 100 Hz. In between these two regions, a transient behavior is observed. In the transient as well as the cycle dependent regions, the R value has a reverse influence, indicating a decrease in crack growth rate with increase in R. For the frequency range tested in this investigation, fractographic results are consistent with the data given in Fig. 10. For R values above 0.8, cycle dependent mechanism is negligible and the crack growth occurs mainly due to time dependent processes.

This relationship between crack growth micromechanisms and stress ratio can be extended to understand the major / minor cycle fatigue crack growth behavior of Inconel 718 shown in Fig. 11, in the form of crack growth rate per major cycle (da/dM) versus minor cycle stress

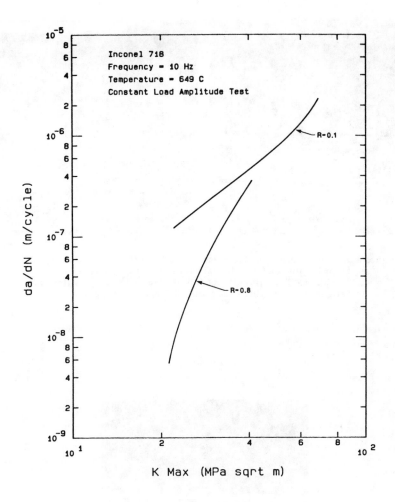

Figure 7. Fatigue crack growth rate (da/dN) versus maximum stress intensity
(K_{max}) plots from constant load amplitude tests run under load
ratios of 0.1 and 0.8. and a frequency of 10 Hz.

intensity range (ΔK_{minor}). In the ΔK_{minor} range above approximately 6
MPa$\sqrt{m}$ there is a continuous increase in the fatigue crack growth rate
with increasing minor cycle stress intensity for all values of ΔK_{major}.
Since the mean stress of the minor cycles (major cycle maximum stress
intensity) remains constant in each test, R_{minor} decreases with

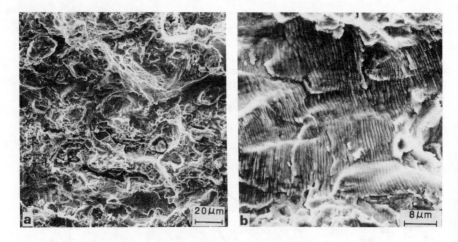

Figure 8. Fracture morphology observed in the Inconel 718 specimen tested
at an R value of 0.1 and maximum stress intensity value of
50 MPa√m. Transgranular mode was observed as in (a) with clear
fatigue striations (b). Crack growth from left to right.

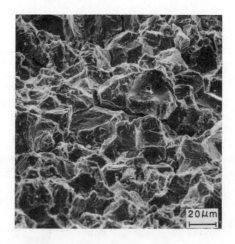

Figure 9. Fracture morphology observed in the Inconel 718 specimen tested
at an R value of 0.8 and a maximum stress intensity of 40 MPa√m.
Intergranular fracture morphology was observed. Crack growth
from left to right.

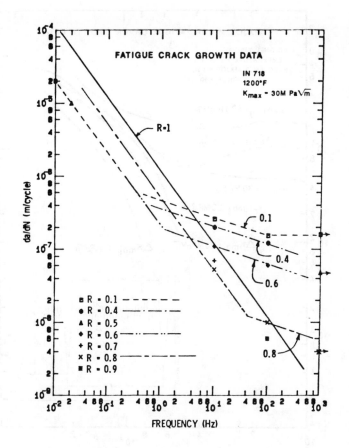

Figure 10. Fatigue crack growth rate (da/dN) versus test frequency (f) plots under a maximum stress intensity (K_{max}) of 30 MPa√m and load ratios (R) between 0.1 and 0 9. Sustained load crack growth data are also shown in solid line (R=1.0).

increasing minor cycle load amplitude. As a result the crack growth behavior can be interpretted as a function of minor cycle R. Increase in the major cycle stress intensity range also results in increased crack growth rates. This is due to a combination of the increased cyclic contribution of the major cycle stress amplitude and the increase

93

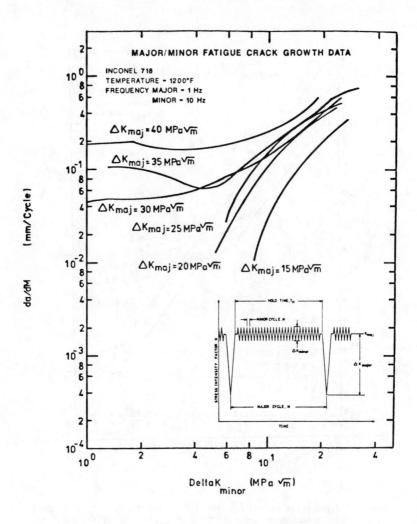

Figure 11. Fatigue crack growth rate (da/dM) versus minor cycle stress
intensity range (K_{minor}) plots for Inconel 718 tested at 649°C
and major cycle stress intensity range between 15 and 40 MPa√m.
Rminor decreases with increasing minor cycle stress intensity
range.

in the stress intensity at the hold time. The latter results in an
increase in the sustained load crack growth contribution and turns out
to be the dominant factor [16]. Finally, a slight decrease in the crack
growth rate is observed with an increasing minor cycle amplitude at the

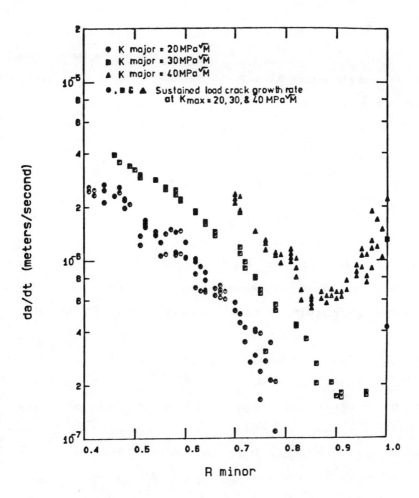

Figure 12. Fatigue crack growth rate (da/dt) versus minor load R value
(R_{minor}) plots for Inconel 718 tested at 649°C.

minor cycle stress intensity range between 2 and 6 MPa$\sqrt{m}$ for tests where
the major cycle stress intensity range is above 35 MPa$\sqrt{m}$. This is
attributed to some type of synergistic effect resulting in a net
decrease in effective crack driving force.

The data of Fig 11 can be replotted as crack grtowth rate (da/dt)
versus R_{minor} for constant values of ΔK_{major} as shown in Fig. 12. Also
shown in Fig. 12 are the points from sustained load crack growth tests,

which are plotted as being equivalent to R = 1.0. Low ΔK_{minor} levels corresponding to high R_{minor} values appear to influence the crack growth behavior mainly due to the major cycle and the hold at the maximum stress intensity.

As shown in Fig. 12, as R_{minor} decreases, that is, as the minor cycle amplitude is increased, a slight decrease in the crack growth rate results. The minimum growth rate occurs in the R_{minor} range of 0.8 to 0.9 for ΔK_{major} of 40 MPa$\sqrt{m}$. For the other two cases of ΔK_{major} equal to 20 and 30 MPa$\sqrt{m}$, it is not as obvious where the minimum occurs because of insufficient data, but a minimum is apparent by observing the values of da/dt for R = 1.0. The relationship between the crack growth mechanisms and stress ratio derived earlier can be used to explain this phenomenon. An interaction between the fatigue and creep contributions to the crack growth process apparently takes place in this region where the R_{minor} value lies between 0.8 and 0.9.

The fractographic analysis of the major / minor test specimens further confirms this observation. The sample tested at the major cycle stress intensity range of 20 MPa $\sqrt{m}$ reveals an intergranular fracture morphology at lower ΔK_{minor} levels (approximately 9 MPa$\sqrt{m}$) as shown in Fig. 13a which corrresponds to an R_{minor} of 0.7. At higher ΔK_{minor} levels, (approximately 20 MPa $\sqrt{m}$), however, a mixed mode fracture morphology indicating a transition to the transgranular mode is observed as indicated in Fig. 13b. The R_{minor} value for this ΔK_{minor} level is 0.4. On the other hand, for the test run with a major cycle stress intensity range of 40 MPa $\sqrt{m}$, the fracture morphology is totally intergranular for the entire minor cycle stress intensity range studied as shown in Fig 14. The minor cycle R, however, is greater than 0.7 for the entire test also indicating a time dependent behavior. As a result, the R_{minor} serves as a tool to indicate the various micromechanisms operating under even complex load spectra.

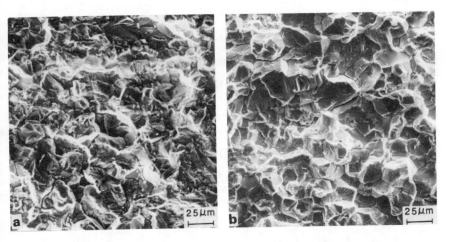

Figure 13. Fracture morphologies observed under major/minor cycle fatigue
of Inconel 718. The major cycle stress intensity range for
the test was 20 MPa√m. (a) Mixed mode (transgranular plus
intergranular) fracture morphologies corresponding to a minor
load R value of 0.4. (b) Intergranular fracture morphology
corresponding to a minor load R value of 0.7. crack growth
from left to right.

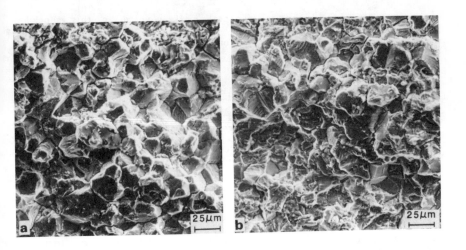

Figure 14. Fracture morphologies observed under major/minor cycle fatigue
of Inconel 718. The major cycle stress intensity range for the
test was 40 MPa√m. Fracture mode was intergranular for the
entire test. The minor load R values corresponding to the above
fractographs were (a) 0.7 and (b) 0.8. Crack growth from left
to right.

Conclusions

Load ratio has a strong influence on the elevated temperature fatigue crack growth behavior of Inconel 718. At high R values the time dependent mechanism operates indicating that mean stress effects are dominant. On the other hand, at low R values the cycle dependent mechanism dominates indicating that the stress amplitude effect dominates. In between these two extremes, a transient behavior occurs where both the mean stress effect and load amplitude effect are contributing. The net result in this region appears to be a reduction in crack growth rate. This transition from time and environment dependent to cycle dependent mechanism occurs at an R value of 0.8 to 0.9. Thus by considering the R value and mean stress level, the crack growth mechanism can be predicted.

Acknowledgements

The authors would like to express their appreciation to R.C. Goodman, N.E. Ashbaugh, and A.M. Brown of the University of Dayton for their contributions to the experimental work covered in this paper. This work was supported by the U.S. Air Force under project 2302P1.

References

1. Hill, R.J., Reimann, W.H., and Ogg, J.S., "A Retirement-for-cause Study of an Engine Turbine disc," AFWAL/ML Report.

2. Nicholas, T., Laflen, J.H., and VanStone, R.H., "A damage Tolerant design Approach to Turbine Engine Life Prediction," AFWAL/ML Report, Wright-Patterson AFB, 1985.

3. Nicholas, T., and Larsen, J.M., "Life Prediction for Turbine Engine Components," Proceedings of the 27th Sagamore Army Materials Research Conference on 'Fatigue: Environment and Temperature Effects', 1980, pp. 353.

4. Nicholas, T., Weerasooriya, T., and Ashbaugh, N.E., "A Model for Creep/Fatigue Interactions in Alloy 718," Fracture Mechanics: Sixteenth Symposium, ASTM STP 868, 1985, pp. 167.

5. Clavel, M., and Pineau, A., "Frequency and Waveform Effects on the Fatigue Crack Growth Behavior of Alloy 718 at 298°K and 823°K, Metallurgical Transactions, vol. 9A, pp. 471, 1978.

6. Weerasooriya, T., "Effect of Frequency on Fatigue Crack Growth Rate of Inconel 718 at High Temperatures," AFWAL-TR-87-4038, Wright-Patterson Air Force Base, June, 1987.

7. Scarlin, R.B., "Effects of Loading Frequency and Environment on High Temperature Fatigue Crack Growth in Nickel Base Alloys," Fracture 1977, Vol. 2, pp. 849, Proceedings of ICFA, Waterloo, Canada, June 1979.

8. Solomon, H.D, and Coffin, L.F., Jr., "Effects of Frequency and Environment on Fatigue Crack Growth in A-286 at 100°F," ASTM STP 520, 1973, pp. 112.

9. Larsen, J.M., and Nicholas, T., in Fracture Mechanics: Fourteenth Symposium - Vol. II : Testing and Applications, ASTM STP 791, 1983, pp. II 536.

10. Shahinian, P., and Sadananda, K., "Effects of Stress Ratio and Hold Time on Fatigue Crack Growth in Alloy 718," Journal of Engineering Materials and Technology, July 1979, vol. 101, pp. 224.

11. Wallace, R. M., Annis, C. G., and Sims, D. L., "Application of Fracture Mechanics at Elevated Temperatures," AFML-TR 76-176 Part II, Air Force Materials Laboratory, April, 1977.

12. James, L.A., "The Effect of Stress Ratio Upon the Elevated Temperature Fatigue Crack Growth Behavior of Several Reactor Structural Materials," HEDL-TME 75-20, Hanford Engineering Development Laboratory, Feb. 1975.

13. Blades, N.A.J., Plumbridge, W.J., and Sidney, D., "High Temperature Fatigue Crack Propagation in the Aluminum Alloy RR58," Materials Science and Engineering, vol. 26, 1976, pp. 195.

14. Corwin, W.R., Booker, B.L.P., and Brinkman, C.R., "Fatigue Crack Propagation in 2 1/4 Cr - 1 Mo Steel," ORNL-5355, Oak Ridge National Laboratory, Feb. 1978.

15. Goodman, R.C., and Brown, A.M., "High Frequency Fatigue of Turbine Blade Materials," Report #AFWAL-TR-82-4151, Materials Laboratory, Wright-Patterson Air Force Base, October 1982.

16. Venkataraman, S., Nicholas, T., and Ashbaugh, N. E., "Micromechanisms of Major / Minor Cycle Fatigue Crack Growth in Inconel 718," Fractography of Modern Engineering Materials: Composites and Metals, ASTM STP 948, 1987.

6. Nicholas, T., "Effect of Frequency on Fatigue Crack Growth Rate of Inconel 718 at High Temperature," AFWAL-TR-83-XXX, Air Force Wright-Patterson AF Base, June 1983.

7. Coffin, L.F., "Fatigue at Elevated Temperature and Environment in High Temperature Fatigue Classification," *Base Alloys*, *Fracture 1977*, Vol. 1, pp. xxx, Proceedings of 4th International Congress, June 1977.

8. Solomon, H.D. and Coffin, L.F., "Effects of Frequency and Environment on Fatigue Crack Growth in A286 at 1100°F," *ASTM STP* 520, 1973, pp. xxx.

9. Jaske, C.E. and Nicholas, T., in Fracture Mechanics, Fourteenth Symposium, *Vol. II: Testing and Applications*, *ASTM STP* 791, 1984, pp. II-xxx.

10. Mukherjee, P. and Shahinian, P., "Cyclic Crack Growth at Elevated Temperature in ..., Fatigue Crack Growth Behavior in Alloy 718," *Journal of Engineering Materials and Technology*, July 1973, Vol. 101, pp. xxx.

11. Pelloux, R.M., Ahmad, M.O., and Shahdi, ..., "Application of Mechanistic Models at Elevated Temperatures," *AFML-TR-xx*, ... Right AI Force Materials ... Laboratory, April 1979.

12. Jones, D.L., "The Plateau Stress Ratio Upon the Elevated Temperature Fatigue Crack Growth Behavior of ... Metal Fatigue Behavior, Materials," *HEDL-TME 76-28*, Washington Engineering Development Laboratory, Dec. 1978.

13. Fleck, W.G., Bill, R.C., and Shahinian, P., "Effect of ... on the Fatigue Crack Propagation in the Aluminum Alloy Rb-xxx," *Metals Science and Engineering*, Vol. xx, 1973, pp. xxx.

14. Brown, W.F. Jr. and Srawley, J.E., Plane Strain Crack ... and ..., *Manual of KIc of High Strength Metallic Materials*, ... for 1978.

15. Goodman, R.C. and Brown, A.M., "High Frequency Fatigue of Turbine Blade Materials," Report No. AFWAL-TR-..., Materials Laboratory, Wright-Patterson Air Force Base, October 1981.

16. Venkateswaran, ..., Tobias, ..., and Ashbaugh, N.E., "Mechanisms of High Strain-Cycle Fatigue Crack Growth in Inconel 718," ... Proceedings of 2nd Engineering Mechanics ... Composites and Metals, *ASTM STP* XXXXXX.

FREQUENCY AND ENVIRONMENT EFFECTS

ON CRACK GROWTH IN INCONEL 718

Tusit Weerasooriya[*] and Srivathsan Venkataraman[**]

* University of Dayton Research Institute
Dayton, Ohio 45469

** Systran Corporation, Dayton, Ohio 45432

ABSTRACT

The fatigue crack growth rate behavior in vacuum of Inconel 718 as a function of frequency at a constant maximum stress-intensity factor (K_{max}) is presented and compared with the growth rate data in air from previous work. The range of frequencies covered is 0.001-1 Hz. Most of the experiments were conducted at R = 0.1 and 0.5, and also a limited number was conducted at R = 0.8. It is found in vacuum that the fatigue crack growth rates as a function of frequency can be represented by three distinct crack growth regimes, as observed in air. These regimes represent three different types of crack growth behavior: cycle-dependent, mixed, and time-dependent. Fracture surface observations confirm these three different modes of damage. Comparison of the results in vacuum with that in air shows that the air environment has a detrimental influence on the crack growth in the time-dependent and mixed regimes, but not in the cycle-dependent regime. Also, the mechanism of intergranular crack growth in the time-dependent regime is altered from environmental degradation to creep cavitation of the grain boundaries ahead of the crack tip by the change of the environment from air to vacuum.

INTRODUCTION

In developing life prediction methodology for high temperature components, a knowledge and an understanding of the crack growth processes that result from the interaction of the parameters such as temperature, frequency, R-value, K_{max} and environment are necessary. It is also very essential to know the mechanism of crack growth for a given combination of these parameters, so that the correct models are selected in describing the fatigue crack growth rate (FCGR) behavior in life prediction.

In general, subcritical crack growth occurs by a cycle-dependent process, a time-dependent process, or by a combination of these two.[1],[2],[3] The process that dominates the FCGR depends on the following parameters: temperature (T), frequency (f), maximum stress intensity (K_{max}), stress ratio (R), environment, and material. For example, an increase of temperature or R-value or a decrease of frequency contributes to the increase of FCGR by increasing the time-dependent contribution to the crack growth mechanism. The presence of an aggressive environment compared to an inert environment, such as a vacuum, also increases the crack growth rate by promoting the time-dependent environmental attack at the crack tip. In a vacuum, another type of time-dependent process such as creep damage may be dominant over the environmental degradation.

In previous work, dependence of FCGR in air on frequency, temperature, R and K_{max} was studied in detail[1]. The environment was observed to play a dominant role in the time-dependent regime of crack growth. This paper describes a study of FCGR in vacuum as a function of frequency and R. The resulting crack growth rates and mechanisms are compared with that observed previously in air.

EXPERIMENTAL PROCEDURE

A series of fatigue crack growth tests was conducted at a constant K_{max} on a Nickel-base superalloy Inconel 718 in a vacuum. A standard (ASTM E647) C(T) specimen of width, W, 20 mm and thickness, B, 5 mm was used in the test series. All specimens were fabricated from the same heat of material. The chemical composition and the heat treatment of Inconel 718 material have been previously reported[1].

An automated test system based on a micro-computer and an electro-hydraulic servo-controlled test machine was employed in conducting these vacuum experiments. Using a turbomolecular pump, a vacuum of less than 10^{-6} was maintained during the testing. The test specimen was heated with molybdenum resistance elements with tantalum heat shields, and the temperature at a point on the specimen was controlled to a chosen value within $\pm 2^{\circ}C$. The temperature variation along the crack path was noted to be less than $5^{\circ}C$.

During the vacuum testing, crack length was determined indirectly at periodic cycle intervals using the compliance method[1] with the aid of the micro-computer and an extensometer with quartz extension arms mounted outside the furnace. The cycle intervals were chosen to give crack length increments less than 0.025 mm. After the crack length was determined, the load that was necessary to keep the K_{max} constant was determined by the computer software using the stress intensity relationship given in ASTM E647. If this load differed by more than 0.1 % from the current load, the new load was applied to the specimen.

All crack growth tests were conducted with a triangular loading waveform under a constant stress intensity value of 40 MPa.m$^{1/2}$, which lies in the mid-power-law region of FCGR vs. K_{max} curve.[2],[4] These tests were conducted at a constant temperature of 650°C and R = 0.1, 0.5 and 0.8, in a range of frequencies between 0.001 and 10 Hz.

In the first set of tests, R was held at 0.1, and the crack was grown in 0.75 mm. segments, for each selected frequency. Using the linear regression representation of the crack length-number of cycles data over the region of cycles where a constant growth rate was observed, the rate was computed at the end of each segment. For all the test frequencies, a steady state growth rate was achieved within a 0.25 mm. extension of the crack. Using the same procedure used at R = 0.1, the growth rates were also obtained at the other two R-values.

RESULTS AND DISCUSSION

Crack Growth in Air

Previous work in air at 650°C[1],[3] exhibited in Figure 1 showed the existence of three regimes of crack growth behavior: an environmentally controlled time-dependent, a fatigue controlled cycle-dependent, and a mixed (or environmentally enhanced cycle-dependent). These regimes directly correspond to the observed three different fracture surfaces shown in Figure 2 and the corresponding mechanisms of crack growth depicted in the upper half of the Figure 3. In the first regime, crack growth occurred intergranularly due to the embrittlement of the constituents along the grain boundaries. In the second regime, cracks grew transgranularly aided by the plastic slip process (dislocation glide along the slip planes) at the crack tip region. In the third growth regime, crack growth occurred by a mixture of these two methods .

Crack Growth in Vacuum

FCGR data that were obtained for different frequencies in vacuum at 650°C and at R = 0.1, 0.5 and 0.8, at a K_{max} value of 40 MPa.m$^{1/2}$ are shown in Figure 4 (half-filled data points with the trend represented by dashed lines). In this figure, corresponding crack growth rate data at 650°C and 40 MPa.m$^{1/2}$ in air from the previous work[1] are also given (filled data points with the trend represented by continuous lines).

For R = 0.1, FCGR is approximately constant for the frequencies greater than 0.02 Hz, but begins to increase with decreasing frequency below 0.02 Hz. For R = 0.5, frequency dependence of FCGR starts at a higher frequency, 0.1 Hz. As in R = 0.1, a constant growth rate is observed above this frequency. At R = 0.8, a frequency dependence is shown by all the observed data. Due to the lack of time, at very low growth rates, FCGR data at R = 0.8 were not generated in the fully cycle-dependent regime.

In summary, FCGR data that were obtained in vacuum show frequency independent (cycle-dependent) and frequency dependent regimes of crack growth. In the frequency dependent regime, growth is not fully time-dependent because here the growth rate with respect to the time is not a constant as in air[1]. In vacuum, fully time-dependent regime would start at a frequency even lower than 0.001 Hz or at a R-value higher than 0.8 (approaching the sustained loading conditions). The frequency at which the frequency-dependence (or mixed) starts is a function of the R-value and increases with R.

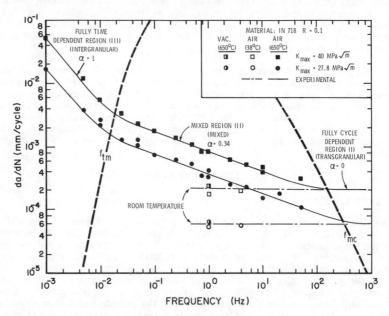

Figure 1 - Fatigue crack growth rate (da/dN) for Inconel 718 as a function of frequency at R = 0.1, temperature = 650°C at two different K_{max} values, 27.8 and 40 MPa.m$^{1/2}$. Also data in vacuum at 1 Hz and room temperature air data are given.[1]

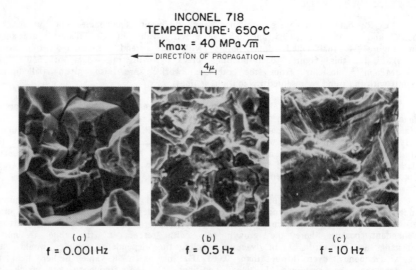

INCONEL 718
TEMPERATURE: 650°C
K_{max} = 40 MPa√m

Figure 2 - Scanning electron fractographs of typical fracture surfaces for (a) time-dependent (0.001 Hz) (b) mixed (0.5 Hz), and (c) cycle-dependent (10 Hz) regions. These tests have been conducted in air at R = 0.1, T = 650°C, and K_{max} = 40 MPa.m$^{1/2}$.[1]

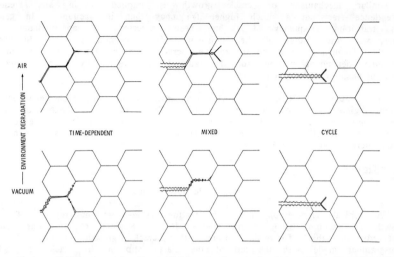

Figure 3 - Schematic of the crack growth mechanisms for time-dependent, mixed and cycle-dependent crack growth regimes in aggressive (air) and non aggressive (vacuum) environments.

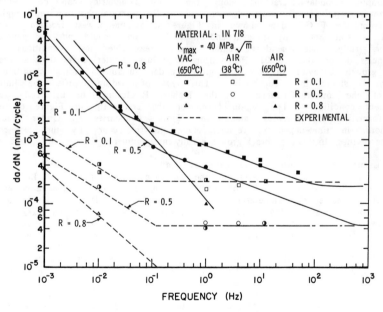

Figure 4 - Fatigue crack growth rate, (da/dN) for Inconel 718, in air and vacuum, as a function of frequency at K_{max} = 40 MPa.m$^{1/2}$ and T = 650°C for three different R values, 0.1, 0.5, and 0.8. Also some room temperature air data are given for R = 0.1 and 0.5. Air data shown in this figure were taken from previous work (filled data points with continuous trend lines: 650oC, and un-filled data points: room temperature).[1]

In the cycle-dependent regime, the frequency independent value of crack growth rate is the same for both air and vacuum at different R-values, thus, a similar mechanism of crack growth is implied. In air, frequency dependence starts at a much higher frequency than in vacuum. In general, this transition frequency, f_{mc}, has been increased by approximately 1000 times, with the change of the environment from vacuum to air. In contrast to the frequency independent regime, crack growth rates in air in the frequency dependent regimes for the same R are higher by approximately two orders of magnitude at a given frequency.

In vacuum, FCGR increases with decreasing R in both cycle-dependent and mixed regimes as also observed in air. Thus, FCGR is controlled by the stress intensity range. In contrast, FCGR in air in the time-dependent regime is controlled by either the maximum or the mean stress intensity-FCGR increases with increasing R.[1]

Crack Growth Mechanisms

Fractographic Observations

A typical fractograph taken from the cycle-dependent regime of crack growth in vacuum is given in Figure 5(a). Regions of striations can be observed in this figure. Here, the crack growth occurs in fully transgranular mode with the aid of the plastic slip at the crack tip. These fracture features are identical to that observed in air (Figure 2(c)) under cycle-dependent conditions. A typical fractograph taken from the frequency dependent mixed regime of crack growth is given in Figure 5(b). In this fractograph, secondary cracking along the grain boundaries can be observed, but the majority of crack growth is still transgranular (mixed regime). Also from the FCGR data shown in Figure 4, at the lowest frequency, crack growth is still not fully time-dependent. Since data could not be generated in the fully time-dependent region under reasonable test conditions, i.e., frequencies lower than 0.001 Hz and/or R values greater than 0.8, a typical crack growth behavior of a crack grown under sustained loading condition in vacuum is shown in Figure 6.[5] This type of loading typifies the limiting case of the lowest frequency or the highest R leading to the fully time-dependent condition. The region ahead of the crack tip in Figure 6 shows the formation of cavities along the grain boundaries. Under vacuum conditions, in time-dependent regime, crack growth occurs by the coalescence of these cavities. In this case, crack growth is fully intergranular.

Possible crack growth mechanisms leading to the observed fracture morphology in vacuum are given in the lower half of Figure 3. In the air where the material ahead of the crack tip is degraded by the environment, with the increase of temperature or R, or with the decrease of frequency, crack growth changes from the cycle-dependent (transgranular fracture), to a mixed, and finally to a time-dependent mechanism (intergranular fracture).[1],[3] In the other extreme, when the environmental damage is not possible (in vacuum), with the increase of R, or the decrease of frequency, crack growth morphology changes from the transgranular, to the mixed, and finally to the intergranular as in air, but the time-dependent (from tests under sustained load) intergranular fracture occurs due to a pure creep cavitation mechanism. The resulted three different mechanisms of crack growth (due to the change of frequency, R, temperature and environment) that are depicted in Figure 3 are: cycle-dependent transgranular, time-dependent intergranular environmental dominant, and time-dependent intergranular creep dominant. These mechanisms are discussed in detail in the following sections.

INCONEL 718
T = 650°C ; K_{max} = 40 MPa $\sqrt{m}$
ENVIRONMENT: VACUUM
◄——— DIRECTION OF PROPAGATION ———

10μ

(a)   (b)
f = 1 Hz : R = 0.1   f = 0.001 Hz : R = 0.8

Figure 5 - Scanning electron fractographs of typical fracture surfaces for (a) cycle-dependent (f = 1 Hz and R = 0.1) and (b) mixed (f = 0.001 Hz and R = 0.8) regimes of crack growth in vacuum. These tests were conducted in air at T = 650°C, and K_{max} = 40 MPa.m$^{1/2}$.

10μ

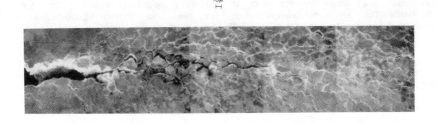

Figure 6 - Crack tip region details of a time-dependent (sustained load) crack in vacuum at 80 MPa.m$^{1/2}$. This test has been conducted at 650°C.[5]

107

Cycle-dependent Crack Growth

As shown earlier, in the cycle-dependent crack growth regime, FCGR is independent of the frequency. In air and vacuum, similar FCGR data and fracture surface morphologies were observed. Therefore, in this regime there is no environmental contribution to the elevated temperature fatigue crack growth behavior. A cycle-dependent process, primarily the formation of striations (as exhibited in Figures 2c and 5a) by the glide of dislocations along the slip planes at the crack tip, is the dominant mechanism of crack growth leading to a transgranular fracture morphology.

Time-dependent Crack Growth

In air and vacuum, time-dependent FCG occurs by intergranular fracture. Depending on the aggressiveness of the environment, the mechanism of intergranular fracture changes from environmentally enhanced to a pure creep cavitation type of mechanism. The observed decrease in FCGR by a factor of 100 for a given frequency in the time-dependent regime of crack growth for R = 0.1, T = 650°C and Kmax = 40 MPa.m$^{1/2}$ can be explained by this change in the growth mechanism - from environmentally dominant to pure creep dominant.

Environmentally Dominant Crack Growth. From previous work[1] in the fully time-dependent region in air, crack growth occurred along the grain boundaries. Among the factors that could account for the time-dependent mechanisms are oxidation damage and creep damage along the grain boundaries. The environment appeared to be the dominant mechanism of damage under time-dependent conditions in air: there was evidence for environmental degradation on the fracture surface - cracks were covered with oxides at lower frequencies. Also there was no evidence of any creep cavitation along the fracture surface grain boundaries or along the grain boundaries ahead of the crack tip. In general, oxygen can diffuse through grain boundaries faster than through the bulk material. Environmental embrittlement of nickel-base superalloys at high temperature by oxygen penetration along the grain boundaries has been postulated in detail by Woodford and Bricknell.[6] In nickel-base superalloys, embrittlement is due to the penetration of gaseous species, primarily oxygen, along the grain boundaries; thus, weakening them either by formation of brittle phases or by removal of strengthening phases. Therefore, apparent activation energy value calculated from crack growth data in the time-dependent regime (245 kJ/mol) represents the activation energy value for the rate determining step of the kinetics of fracture along the grain boundaries at the crack tip.[3]

Creep Dominant Crack Growth. Even though in this study we were unable to generate crack growth data in the time-dependent region, from the sustained load data in vacuum, time-dependent crack growth occurred along the grain boundaries. Creep damage normally occurs by grain boundary and triple point cavitation.[7],[8] Cavitation damage arises primarily by grain boundary sliding and diffusional cavity growth mechanisms.[9] As discussed earlier and shown in Figure 6, examination of the region ahead of the crack tip shows the formation of creep cavities along the grain boundaries. Floreen[10] has also reported grain boundary cavities on the fracture surface of Inconel 718 tested under sustained load in inert helium. Crack growth occurs by the coalescence of the cavities at the crack tip region. For the rate limiting step of the kinetics of this type of crack growth mechanism (time-dependent in vacuum), an apparent activation energy value of 97 kJ/mol is reported by Floreen.[4] In contrast, the apparent activation energy value for the environmentally dominant crack growth mechanism is much higher.

CONCLUSIONS

The frequency dependence of the linear mid-range fatigue crack growth rate behavior exhibited in air and vacuum at 650°C by Inconel 718 can be divided into three crack growth regimes -- fully cycle-dependent, mixed, and fully time-dependent damage -- irrespective of stress ratio, maximum stress-intensity factor, and frequency. Micromechanisms of fatigue crack growth which were deducted from fracture surface analysis support the existence of these three regimes of growth. Although the growth rates were the same in the frequency independent regimes both in air and vacuum, in the frequency dependent regime, FCGR could be as much as two orders of magnitude higher in air for the same R and frequency. Environment has a strong influence on the time-dependent crack growth rates, but has little influence on the cycle-dependent crack growth rates. The mechanism of crack growth changes in the presence of the aggressive air environment from cavitation damage (creep) to oxide embrittlement damage in the time-dependent crack growth regime. In the cycle-dependent regime, the presence of the aggressive environment does not change the mechanism of crack growth - the formation of ductile striation, due to the slip by dislocation glide along the slip planes at the crack tip.

ACKNOWLEDGMENT

This work was supported by the Materials Laboratory at Wright-Patterson Air Force Base and was conducted in the laboratories of the Metals and Ceramics Division of the Materials Laboratory.

109

REFERENCES

1. T. Weerasooriya, "Effect of Frequency on Fatigue Crack Growth Rate of Inconel 718 at High Temperature," Accepted for Publication in ASTM STP 969 of the Proceedings of the 19th National Symposium on Fracture Mechanics. Also available as AFWAL-TR-87-4038, Submitted for Review to Air Force Wright Aeronautical Laboratories, Air Force Materials Laboratory, Wright-Patterson Air Force Base, Ohio, June 1987.

2. K. Sadananda and P. Shahinian, "High Temperature Time-Dependent Crack Growth," Micro and Macro Mechanisms of Crack Growth, Edited by Sadananda, K., Rath, B. B. and Michel, D. J., A Publication of The Metallurgical Society of AIME, 1982, pp. 119-130.

3. T. Weerasooriya, "Temperature and Frequency Effects on Fatigue Crack Growth Rate of Inconel 718," To be published.

4. S. Floreen, "Effects of Environment on Intermediate Temperature Crack Growth in Superalloys," Micro and Macro Mechanisms of Crack Growth, Edited by K. Sadananda, B. B. Rath and D. J. Michel, A Publication of The Metallurgical Society of AIME, 1982, pp. 177-184.

5. M. Stucke, M. Kobaib, B. Majumdar and T. Nicholas, "Environment Aspects in Creep Crack Growth in a Nickel-base Superalloy," Proceedings of ICF-6, India, 1984, pp. 3967-3975.

6. R. H. Bricknell and D. A. Woodford, " The Embrittlement of Nickel Following High Temperature Air Exposure", Met. Trans., Vol. 12A, 1981, p. 425.

7. R. Raj and M. F. Ashby, "Intergranular Fracture at Elevated Temperature", Acta Met., 21, 1975, p.1625.

8. R. J. Fields, T. Weerasooriya and M. F. Ashby, "Fracture-Mechanisms in Pure Iron, Two Austenitic Steels, and One Ferritic Steel", Met. Trans., Vol. 11A, 1980, p. 333.

9. R. Raj, "Mechanisms of Creep-Fatigue Interaction," Flow and Fracture at Elevated Temperature, Edited by Rishi Raj, American Society for Metals, Ohio, 1984, pp.215-249.

10. S. Floreen, "High Temperature Crack Growth Structure-Property Relationships in Nickel-base Superalloys," Creep-Fatigue-Environment Interactions, Edited by Pelloux, R. M. and Stoloff, N. S., A Publication of The Metallurgical Society of AIME, 1980, pp.112-128.

Isothermal and Thermal-Mechanical Fatigue

Fatigue and Thermal Fatigue of Pb-Sn Solder Joints

D. Frear, D. Grivas, M. McCormack, D. Tribula, J. W. Morris, Jr.

Department of Materials Science and Mineral Engineering
University of California, Berkeley
Berkeley, CA 94720

ABSTRACT

This paper presents a fundamental investigation of the fatigue and thermal fatigue characteristics, with an emphasis on the microstructural development during fatigue, of Sn-Pb solder joints. Fatigue tests were performed in simple shear on both 60Sn-40Pb and 5Sn-95Pb solder joints. Isothermal fatigue tests show increasing fatigue life of 60Sn-40Pb solder joints with decreasing strain and temperature. In contrast, such behavior was not observed in the isothermal fatigue of 5Sn-95Pb solder joints. Thermal fatigue results on 60Sn-40Pb solder cycled between -55°C and 125°C show that a coarsened region develops in the center of the joint. Both Pb-rich and Sn-rich phases coarsen, and cracks form within these coarsened regions. The failure mode of 60Sn-40Pb solder joints in thermal and isothermal fatigue is similar: cracks form intergranularly through the Sn-rich phase or along Sn/Pb interphase boundaries. Extensive cracking is found throughout the 5Sn-95Pb joint for both thermal and isothermal fatigue. In thermal fatigue the 5Sn-95Pb solder joints failed after fewer cycles than 60Sn-40Pb.

INTRODUCTION

Sn-Pb solder joints are used extensively in the electronics industry as an electrical/mechanical interconnection in electronic packages. A major requirement of the solder joints is that they absorb strains arising from thermal expansion mismatch of unlike materials within the package. For instance, the mismatch between a polyimide circuit board and a leadless ceramic chip carrier (or between a Si chip and the ceramic carrier) produces a shearing strain on the solder joints when the package encounters thermal fluctuations. These temperature fluctuations arise from both enviromental and power cycling. The large fluctuating shear strains create cracks in the solder joint and lead to early catastrophic failures (1-3). Therefore, there is a significant need to better understand the fatigue and thermal fatigue properties of the solders used in electronic packaging.

The purpose of this study is to investigate the fatigue and thermal fatigue characteristics of Sn-Pb solder joints and to relate these results to the microstructural mechanisms that determine fatigue life. Previous work on the Sn-Pb system in conditions of isothermal fatigue (4-9), fatigue at constant strain and temperature, were performed on bulk solder or in various designs of solder joint test specimens. Thermal fatigue tests were performed directly on the solder joints in electronic packages (10-24). However, a correlation between the two fatigue conditions is lacking. This paper presents results of a study of two commonly used Sn-Pb alloys, 60Sn-40Pb and 5Sn-95Pb, soldered to Cu surfaces. These solders were studied in a joint configuration in conditions of thermal fatigue and isothermal fatigue with an emphasis on the microstructural development during fatigue.

EXPERIMENTAL PROCEDURE

Special techniques for the manufacture and fatigue testing of solder joints have been developed in this laboratory and are briefly described here. A more detailed discussion is presented elsewhere (25).

The double shear specimen used to test solder joints in isothermal fatigue is shown in Figure 1a. The two solder joints between the holes experience simple shear on loading along the long axis. The specimen used to test solder joints in thermal fatigue is shown in Figure 1b.

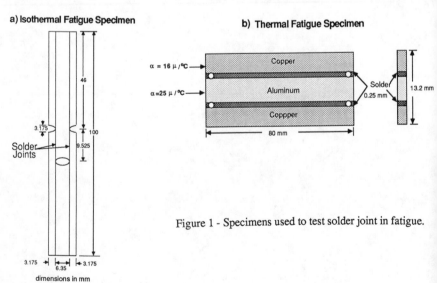

Figure 1 - Specimens used to test solder joint in fatigue.

The specimen consists of an Al plate ($\alpha = 25$ ppm/°C) sandwiched between two Cu plates ($\alpha = 16.6$ ppm/°C). The Al was plated with 0.05 mm (0.002 in) of Ni to act as a diffusion barrier, and 0.025 mm (0.001 in) of Cu to give the solder joint a similar interface on each side. On thermal cycling, the solder joints between the Cu and Al plates undergo simple shear deformation.

To manufacture solder joint specimens, an assembly of plates were bolted together with the appropriate spacers to form a gap, Figure 2. The assembly was submerged in a molten solder bath, in vacuum, and then cooled. Individual specimens were then cut from the assembly. This insures microstructural consistency between test specimens within the same block. Each block yields 8-10 specimens. The solders were made using 99.99 pure Sn and 99.9 pure Pb.

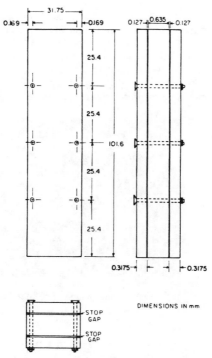

Figure 2 - Assembly of plates used to manufacture solder test specimens.

Isothermal tests were performed on a digitally controlled loadframe. Tests were performed at -55°C (ethyl alchohol cooled by liquid nitrogen), 0°C (ice water), room temperature, and 125°C (quench oil). A through-cracked joint was consistently associated with a 30% decay in cyclic load amplitude. This 30% decay was then used to define the number of cycles to failure. A constant strain rate of 0.05 mm/min was used. The cycling frequencies for the total strains imposed were as follows:

$$35\% \text{ strain} = 8.4 \text{ cycles/hour,}$$
$$30\% \text{ strain} = 9.8 \text{ cycles/hour,}$$
$$25\% \text{ strain} = 11.8 \text{ cycles/hour,}$$
$$20\% \text{ strain} = 14.8 \text{ cycles/hour.}$$

Thermal cycling tests were performed by cycling the specimens between two thermal baths, one at -55°C the other at 125°C. The hold time in each bath was 5 minutes with a transfer time of 30 seconds. The use of liquid thermal baths has two advantages: 1) a rapid heat transfer between the specimen and the bath, 2) an inert atmosphere that eliminates oxidation, which is known to decrease the fatigue life of solder (10).

In the Scanning Electron Microscope (SEM), contrast of the Pb-rich phase is light and the Sn-rich phase dark, in optical micrographs the Pb-rich phase is dark and the Sn-rich phase light.

RESULTS

Initial Microstructure

The initial microstructure of the 60Sn-40Pb on Cu solder joint is shown in Figure 3. The solder consists of globular Pb-rich and Sn-rich phases mixed in with Pb-rich dendrites and eutectic lamellae. The interfacial structure, Figure 4, is a two phase intermetallic Cu_6Sn_5 adjacent to the solder and Cu_3Sn adjacent to the Cu. In the bulk of the solder intermetallic Cu_6Sn_5 whiskers are also observed (26), Figure 5. The microstructure of 95Pb-5Sn is shown in Figure 6 and consists of a Pb-rich matrix with a distribution of ß-Sn precipitates (27-28). A Cu_3Sn intermetallic forms at the interface between the 5Sn-95Pb and Cu (29), Figure 7.

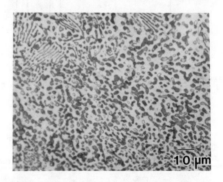

Figure 3 - Optical micrograph of 60Sn-40Pb solder .

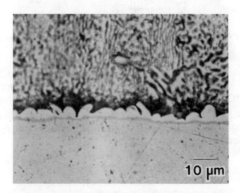

Figure 4 - Optical micrograph of Cu_6Sn_5 and Cu_3Sn intermetallics at interface of Cu and 60Sn-40Pb solder.

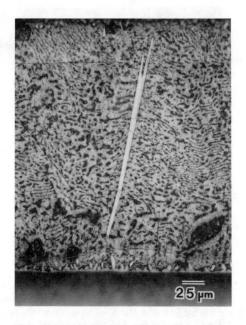

Figure 5 - Optical micrograph showing Cu_6Sn_5 whisker in the solder joint.

Figure 6 - Transmission Electron Micrograph of ß-Sn precipitates in 5Sn-95Pb solder.

Figure 7 - Optical micrograph of Cu$_3$Sn interfacial intermetallic between
Cu and 95Pb-5Sn solder.

Isothermal Fatigue

Isothermal fatigue tests were performed on solder joints in shear to investigate the function
of temperature and total amount of shear strain on the microstructure and failure characteristics
of both 60Sn-40Pb and 95Pb-5Sn solder joints. The results of the fatigue tests on 60Sn-40Pb
joints are shown in Table I. A plot of this data in percent total strain vs. cycles to failure with
varying testing temperature is shown in Figure 8. The number of cycles to failure increases
with decreasing temperature and decreasing strain. The effect of strain on cycles to failure is not
strong at 125°C. Figure 9 shows a failed 60Sn-40Pb solder joint, in cross section, cyclically
strained at 35% and 125°C. Extensive cracking is found along the grain boundaries in the Sn-
rich phase and at the interphase boundaries. Little cracking is observed to occur within the Pb-
rich phase. The microstructure of 60Sn-40Pb solder joints tested at lower temperatures and
smaller strains show cracking through the Sn- rich phase but not to the same extent as at higher
temperatures.

Table I   Cycles to Failure
Isothermal Fatigue
60Sn-40Pb

	125°C	25°C	0°C	-55°C
$\gamma_{35\%}$	10	75	130	1100
$\gamma_{30\%}$	25	125	215	
$\gamma_{25\%}$	60	425	560	
$\gamma_{20\%}$	55	1450	2300	

118

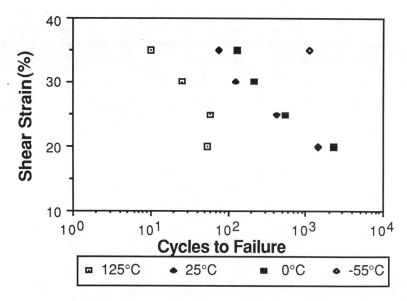

Figure 8 - Plot of percent shear strain vs. number of cycles to failure for
60Sn-40Pb solder joints.

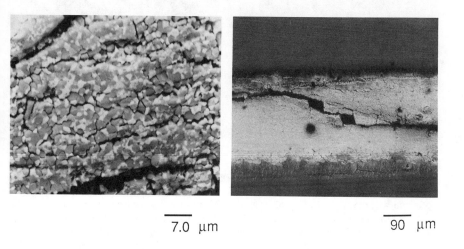

7.0 µm   90 µm

Figure 9 - SEM micrographs 60Sn-40Pb solder joint fatigued at 125°C and
35% shear strain.

119

The results of the isothermal fatigue tests on 5Sn-95Pb solder joints are shown in Table II. This data shows no correlation between temperature, strain and the number of cycles to failure for the 5Sn-95Pb solder joints. Furthermore, the 5Sn-95Pb solder joints fail sooner than 60Sn-40Pb at temperatures below 20°C. At 125°C both solders fail after a similarly short number of cycles. The microstructure of a failed 5Sn-95Pb joint at 125°C and 30% strain is shown in cross section in Figure 10. Extensive cracking throughout the solder is observed both parallel and perpendicular to the direction of shear, and takes on a mosaic structure throughout the joint. This failure mode is observed at all temperatures and strains in the isothermal fatigue tests.

Table II   Cycles to Failure
Isothermal Fatigue
95Pb-5Sn

	125°C	25°C	0°C	-55°C
$\gamma_{35\%}$	45	70	60	45
$\gamma_{30\%}$	70	60	95	35
$\gamma_{25\%}$	80	85	230	
$\gamma_{20\%}$	80	60	200	

500 μm

Figure 10 - SEM micrograph of 5Sn-95Pb solder joint tested at 125°C and 30% shear strain.

Thermal Fatigue -55°C to 125°C

Thermal cycle tests were performed on 60Sn-40Pb solder joined to Cu and Al plates. The joint thickness tested was 0.254 mm (10 mil). The resultant shear strain at the ends of the specimen when cycled between -55°C and 125°C was 14%. A series of cross section optical micrographs taken of different specimens during the fatigue process between -55 °C and 125°C is shown in Figure 11. After 625 cycles a thin coarsened region develops through the center of the solder joint. In this region, both the Sn-rich phase and Pb-rich phase coarsen, Figure 12. Figure 11 reveals that after 1000 cycles cracks are present in the coarsened region, specifically along the grain boundaries of the Sn-rich and the interphase boundaries (Figure 13). In all cases the cracks were found solely in the coarsened region.

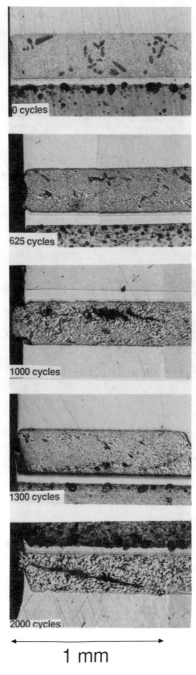

1 mm

Figure 11 - Optical micrographs of 60Sn-40Pb solder joints after thermal fatigue of -55°C to 125°C.

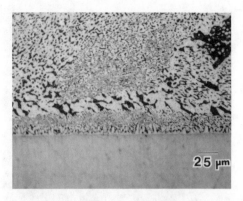

Figure 12 - Optical micrograph of 60Sn-40Pb solder joint revealing coarsened region in the bulk solder where both Pb and Sn phases have coarsened after 625 thermal cycles between -55°C and 125°C.

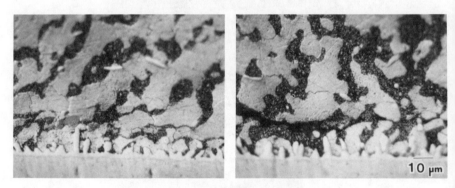

Figure 13 - Optical micrograph of 60Sn-40Pb solder joint after 1000 thermal cycles between -55°C and 125°C showing cracks that occur through the Sn and along the interphase boundaries.

Thermal Fatigue: 95Pb-5Sn

Thermal cycle tests were also performed on 95Pb-5Sn solder joints between Cu and Al plates. The specimens were cycled from -55°C to 125°C. The resultant microstructures are summarized in the SEM micrographs of Figure 14. Cracks form after only 120 thermal cycles. Extensive cracking was found in the joint at 250 and 500 thermal cycles. The cracks run both parallel and perpendicular to the joint. The mosaic failure pattern was observed in both thermal and isothermal fatigue of 5Sn-95Pb. This is clearly apparent in the specimen after 120 thermal cycles, Figure 15.

122

Figure 14 - SEM micrograph of unpolished 5Sn-95Pb solder joint in thermal fatigue
between -55°C to 125°C.

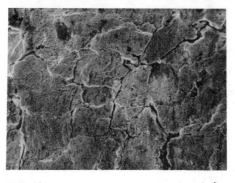

120 cyles   20 μm

Figure 15 - SEM micrograph of unpolished 5Sn-95Pb after 120 thermal cycles
between -55°C to 125°C showing mosiac crack pattern.

DISCUSSION

In isothermal fatigue the number of cycles to failure at room temperature and below is less for 5Sn-95Pb than for 60Sn-40Pb. At 125° the number of cycles to failure is similar for both alloys. The fatigue properties for 60Sn-40Pb deteriorate at 125°C as a consequence of the extensive intergranular cracking in the Sn-rich phase. At lower temperatures many more cycles are required to achieve the same extent of cracking in the Sn-rich phase, and therefore longer isothermal fatigue lives. In contrast, 5Sn-95Pb fails in isothermal fatigue after a short number of cycles for all temperatures tested, and does not display this degradation of fatigue properties.

The thermal cycling tests indicated that 60Sn-40Pb solder has a much longer thermal fatigue life than 5Sn-95Pb. An interesting observation from the thermal fatigue tests of 60Sn-40Pb solder joints is the localized coarsening of both the Sn and Pb-rich phases. In addition to this, cracks were found to form solely within these coarsened regions.

Within a given solder compositio, the isothermal and thermal fatigue failure modes were similar. For the 5Sn-95Pb solder joints, a mosaic failure pattern was found for both tests. In the 60Sn-40Pb cracks form intergranularly within the Sn-rich phase and at the Sn phase/Pb phase boundary.

SUMMARY AND CONCLUSIONS

Isothermal fatigue tests to failure were performed on both 60Sn-40Pb and 5Sn-95Pb solder joints at temperatures between -55°C and 125°C and shear strains from 20% to 35% shear strain. The number of cycles to failure for the 60Sn-40Pb joints increased with decreasing strain and temperature. The 5Sn-95Pb joints showed a fatigue life that could not be correlated with strain and temperature in the ranges tested. At temperatures below 20°C the 5Sn-95Pb solder joints failed after fewer cycles than 60Sn-40Pb joints. At 125°C both solder joints failed after a short number of cycles. Thermal fatigue tests of 5Sn-95Pb and 60Sn-40Pb solder joints cycled from -55° to 125°C were performed. The 60Sn-40Pb joints developed a localized coarsening along the length, through the center of the joint, and through both the Pb-rich and Sn-rich phases. Cracks formed only within these coarsened regions after 1000 thermal cycles. However, failures were observed after only 120 thermal cycles for the 5Sn-95Pb solder joints. The results from thermal fatigue indicate that the 60Sn-40Pb solder joints are more fatigue resistant than 5Sn-95Pb solder joints in cycling from -55° to 125°C .

Within a given solder composition, the same fatigue failure mode was found in both isothermal and thermal fatigue tests. The failure mode of 5Sn-95Pb was a homogeneous mosaic crack pattern throughout the joint. In contrast, in 60Sn-40Pb cracks intergranularly in the Sn-rich phase and along the interphase boundaries.

ACKNOWLEDGEMENTS

This work was supported by the Director, Office of Energy Research, Office of Basic Energy Science, Materials Sciences Division of the U. S. Department of Energy under contract #DE-AC-3-76SF00098.

REFERENCES

1. M. J. Berkebile, "Investigation of Solder Cracking Problems on Printed Circuit Boards," (NASA TMX 53653, September 1967).

2. C. L. Lassen, "Use of Metal Core Substrates for Leadless Chip Carrier Interconnection," Electronic Packaging Production, March (1981), 98-104.

3. C. A. Haper, W. W. Stanley, "Some Critical Mateials Factors in the Application of Leadless Chip Carrier Packages," Electron. Packag. Prod., Aug. (1981), 134-42.

4. R. Wild, "Some Fatigue Properties of Solder and Solder Joints," (IBM Technical Report 74Z-000481, 1975).

5. K. U. Snowden, "The Effect of Atmosphere on the Fatigue of Pb," Acta Met., 12 (1984), 295-303.

6. G. Becker, "Testing and Results Related to the Mechanical Strength of Solder Joints," (Paper presented at the Institute for Interconnecting and packaging Electronic Circuits, Fall Meeting, San Fransisco, CA, September, 1979).

7. L. S. Goldman, "Geometric Optimization of Controlled Collapse Interconnections," IBM J. Res. Dev., 13, (1969), 251-261.

8. N. D. Zommer, D. L. Feucht, R. W. Heckel, Reliability and Thermal Impedance Studies in Soft Soldered Power Transistors," IEEE Trans. Elec. Dev., ED-23, (1976),843-849.

9. C. J. Thwaites, W. B. Hampshire, "Mechanical Strength of Selected Soldered Joints and Bulk Solder Alloys," Welding Res. Supp., October (1976), 323s-329s.

10. J. W. Munford, "The Influence of Several Design and Material Variables on the Propensity for Solder Joint Cracking," IEEE PHP-11, 11 (1975), 296-304.

11. J. R. Taylor, D. J. Pedder, "Joint Strength and Thermal Fatigue in Chip Carrier Assembly," Int. J. Hybrid Microelec., 5 (1982), 209-14.

12. P. M. Hall,"Forces, Moments, and Desplacements Dufing Thermal Chamber Cycling of Leadless Ceramic Chip Carriers Soldered to Printed Boards," IEEE CHMT-7, 7 (1984), 314-327.

13. E. A. Wright, W. M. Wolverton,"The Effect of the Solder Reflow Method and Joint Design on the Thermal Fatigue Life of LCC Solder Joints," Proc. Electronic Components Conf., 34 (1984),149-155.

14. M. C. Delinger, D. W. Becker, "Improved Solder Alloy for Printed Circuit Board Application," Welding J., 57 (1978), 292s-297s.

15. P. M. Hall, T. D. Duddeerar, J. F. Argyle, "Thermal Deformations Observed in Leadless Ceramic Chip Carriers Surface Mounted to Printed Wiring Boards," IEEE CHMT -6, 6 (1983), 544-552.

16. W. Engelmaier, "Fatigue Life fo Leadless Chip Carrier Solder Joints During Power Cycling," IEEE CHMT-6, 6 (1983), 232-237.

17. H. M. Berg, E. L. Hall, "Dissolution Rates and Reliability Effects of Au, Ag, Ni, And Cu in Pb Base Solders," (Proc. Reliability Phys. Symp., (1973), 10-20).

18. D. R. Olsen, H. M. Berg, "Properties of Die Bond Alloys Relating to Thermal Fatigue," IEEE CHMT-2, 2 (1979), 257-263.

19. S. K. Kang, N. D. Zommer, D. L. Feucht, R. W. Heckel, "Thermal Fatigue Failure of Soft Soldered Contacts to Si Power Transistors," IEEE PHP-13, 13 (1977), 318-321.

20. W. Englemaier, "Effects of Power Cycling on LCC mounting Reliability and Technology," Proc. 2nd Ann. Int. Packaging Conf., 2 (1982), 15-22.

21. J. T. Lynch, M. R. Ford, A. Boetti, "The Effect of High Dissipation Components on the Solder Joints of Ceramic Chip Carriers Attached to Thick Film Alumina Oxide Substrates," IEEE CHMT-6, 6 (1983), 237-245.

22. J. F. Burgers, R. O. Carlson, H. H. Glascock, C. A. Neugebauer, H. F. Webster, "Solder Fatigue Problems in Power Packages," IEEE CHMT-7, 7 (1984), 405-410.

23. E. Levine, J. Ordenez, "Analysis of Thermal Cycle Fatigue Damage in Microsocket Solder Joints," IEEE CHMT-4, 4 (1981), 515-519.

24. L. S. Goldmann, R. D. Herdzek, N. G. Koopman, V. C. Marcotte, "Pb-In for Controlled Collapse Chip Joining," IEEE PHP-13, 13 (1977), 194-198.

25. D. Frear, D. Grivas, M. McCormack, D. Tribula, J. W. Morris, Jr., "Fatigue and Thermal Fatigue Testing of Pb-Sn Solder Joints," (to be published Proc. 3rd Ann. ASM Conf. Elec. Packaging Conf., Minneapolis, MN, April, 1987)

26. D. Frear, D. Grivas, J. W. Morris, Jr., "The Effect of Cu_6Sn_5 Whisker Precipitates in Bulk 60Sn-40Pb Solder,", (accepted for publication in J. of Electronic Materials).

27. D. Frear, J. W. Morris, Jr., "Observation of Precipitates in 95Pb-5Sn Solder", Proc. 43rd Ann. Meeting EMSA, 43 (1985), 342-343.

28. J. B. Posthill, D. Frear, J. W. Morris, Jr., "In Situ Observation of Interphase Boundary Motion in 95Pb-5Sn Solder," Proc. 44th Ann. Meeting EMSA, 44 (1986) 410-411.

29. D. Grivas, D. Frear, L. Quan, J. W. Morris, Jr. "The Formation of Cu_3Sn Intermetallic on the Reaction of Cu with 95Pb-5Sn Solder," J. Electronic Materials,15(1986), 355-359.

ISOTHERMAL FATIGUE FAILURE MECHANISMS IN LOW TIN LEAD BASED SOLDER

S. Vaynman and M. E. Fine

Department of Materials Science & Engineering and
Materials Research Center, Northwestern University,
Evanston, IL 60201

D. A. Jeannotte

IBM Corporation, East Fishkill Facility
Hopewell Junction, NY 12533-0999

Abstract

The isothermal fatigue failure process in a low tin lead based solder depends upon experimental conditions. At high frequency ($> 10^{-2}$ Hz) at high total strain range (0.75%) failure mode is mixed transgranular-intergranular, at low total strain range (0.3%) the mode of failure is purely intergranular. Change in failure mode leads to two slopes in the Coffin-Manson plots. Tensile hold time, combined tensile and compressive hold times and decreasing frequency are found to reduce the isothermal fatigue resistance of this solder. The main mode of fracture under these conditions is intergranular.

127

Introduction

Failure of solders due to thermomechanical fatigue is a very serious problem in the electronics and power industries (1,2); however, there is a paucity of data on the fatigue behavior and mechanisms of fatigue failure of solders versus strain range, temperature, frequency and hold time. In the absence of grain boundary embrittlement fatigue at $T < 0.4$ T_m, where T_m is absolute melting temperature, is characterized by a transgranular mode of fracture. At higher temperatures the time-dependent processes such as creep and accelerated crack propagation from environmental reactions are intensified, and as a result there is a transition from a transgranular to intergranular mode of fracture. Since room temperature is approximately equal to 0.5 T_m for lead and lead alloys, an intergranular mode of fracture has been documented in most cases (3,4). However, for 96 wt.% Pb-4 wt.% Sn solder tested at high strains, the deformation was found to be concentrated in a few slip bands where cracks finally formed (5).

J. F. Eckel (6) found a marked influence of frequency on fatigue life of lead; decreasing frequency decreased the fatigue life. A similar frequency effect was found in fatigue of lead with 1% Sb (7), 95Pb-5Sn (3), 60Pb-40Sn (8) solders and other materials (16,18). In many alloys decreasing frequency lead to change the fracture mode from transgranular at high frequency to intergranular at low frequency (11).

While the frequency effect on the fatigue of lead and solders has been studied in some detail, there is little work on the effect of hold times on solder fatigue life. Tensile hold time was found to be very damaging during fatigue of 70In-30Pb solder at room temperature (5). Holds at maximum and minimum temperatures were shown to reduce the thermal fatigue life of 95Pb-5Sn (9) and 50Pb-50In (10) solders. For many metals introduction of a tensile hold time into the cycle lead to the change in the fracture mechanism from transgranular to intergranular (11).

The study of the effects of temperature, strain range, hold time, and frequency on the isothermal fatigue lifetime and failure mode of 96.5Pb-3.5Sn solder between the 5 and 100° C was undertaken to provide data base for design purposes.

Experimental Procedure

This research was done with low tin lead based solder purchased in accordance with an existing vendor-purchaser specification. The composition was determined by the supplier to be 96.5Pb-3.5Sn with less than 0.1% total impurities, which was within the allowable range specified. The impurities were reported as follows: S, P < 5 ppm each; Cu, Bi, Zn, Al < 10 ppm each; Fe, Au, As < 20 ppm each; Ag < 30 ppm; and Sb < 50 ppm.

The specimens were cast in a flat open aluminum mold. After casting, they were machined flat and then homogenized at 175° C for 100 hours. A second anneal at 150° C for 2 hours was done approximately one week before testing to standardize microstructure. A coarse cast microstructure was produced with grain dimensions depending on the orientation of the solder in the mold during crystallization. The grains were elongated in the direction of the specimen width and varied widely from approximately 50 to 500 microns in the short dimension and up to a few millimeters in the long direction. These large grains were divided into subgrains 25-100 micrometers long. Since room temperature is below the solvus line (81° C) for this solder (11), precipitates of tin in the form of lamellae were observed throughout the specimen. The spacing between lamellae was 3 to 15 microns.

128

The specimens were mechanically polished with decreasing grit size down to 1 micrometer. Specimens used for optical or scanning electron microscopic examinations were further electrochemically polished in 70:30 acetic acid-perchloric acid solution at 15-20 volts for 2-3 minutes (12). To avoid etching of specimens along the tin precipitate-matrix interfaces, specimens were electrochemically polished immediately after the 100 hour homogenizing anneal. Each specimen was 6 mm thick and 12 mm wide in the gage section. All mechanical testing was done under total strain control on computerized MTS servo-valve controlled electrohydraulic testing machine. Fatigue tests were done in pull-pull using saw-tooth wave form, i.e., the strain was changed during testing from zero to the maximum value. While the total strain varied from zero to maximum strain, the values for peak tensile and compressive stresses were almost equal due to high plastic strain in solder. In some tests a hold time was introduced at maximum and/or at zero strain. The same type of stress response versus strain was observed in tests with hold time. Therefore, hold time at maximum strain was with a tensile stress while hold time at zero strain was in the presence of a compressive stress. The total strain range held constant during each test was varied from 0.30 to 0.75%.

Heating and cooling of specimens were done through the grips of the testing machine using an apparatus developed by L. Lawson at Northwestern University and described elsewhere (13). Specimens were kept in the grips of testing machine for 2 hours at the testing temperature before the start of cycling. Temperature variation along the gage section of the specimen did not exceed $\pm$ 1° C. Testing was done in air of approximately 50% relative humidity; except to avoid moisture condensation on the specimen during testing at 5° C, dry air was flushed through the specimen chamber.

Replicas of the specimen surfaces for SEM observation taken after predetermined numbers of cycles were coated in vacuum with aluminum and then gold shadowed, giving negative casts of the specimen surfaces (cracks appear as white ridges). The examination was done on a Hitachi 570A or S510 scanning electron microscope using an acceleration voltage of 20 kV. The voltage was increased to 25 kV for microscopy of sections cut from failed specimens.

Testing in the strain controlled mode usually causes cyclic hardening in annealed metals; the stress needed to enforce the maximum strain limit increases with cycling until a saturation stress value is reached. Then, due to microcrack growth, there is a decrease in the value of maximum stress. 96.5Pb-3.5Sn solder exhibited such behavior during fatigue testing. The cycle number corresponding to the maximum value of tensile stress was defined as the number of cycles to failure in this research. This maximum stress was used in research by others (14,15) for initiation of an "engineering crack". As shown later, the specimen is extensively cracked at this definition of failure.

Results and Discussion

1. Continuous Strain Wave Tests

Figure 1 shows the dependence of the fatigue life of 96.5Pb-3.5Sn solder on plastic strain range at temperatures from 5 to 100° C with 0.1 to 2.5 seconds of ramp time. It is evident that the data for each temperature, except possible 5° C where only four data points were taken, cannot be well fit to a single log-log straight line. The data are much better represented by two least square straight lines with a breakpoint at approximately 0.3% of plastic strain.

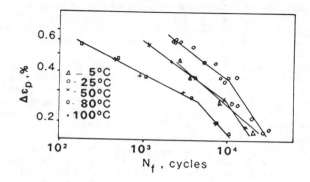

Figure 1 - Log-log plot of the number of cycles to
failure versus plastic strain range. 0.1-2.5 sec
ramp time. No hold time.

In general, cracks could be observed under all testing conditions very
early in the fatigue life for all strain ranges investigated (less than 5%
of the total cycles to failure as defined in this research). Many areas of
specimen were examined and micrographs in this paper are typical. Figures 2
(a replica at N/N_f = 0.04) and 3 (a surface of failed specimen) show that
fatigue of this solder at 0.75% total strain range (2.5 seconds of ramp time)
is due to intergranular cracking as well as transgranular cracking along slip
bands.

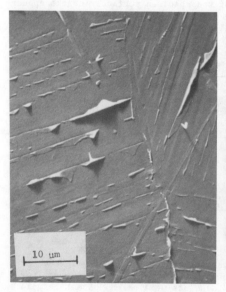

Figure 2 - Scanning electron micrograph of surface replica. Total
strain range 0.75%. 25°C. Ramp time
2.5 sec. No hold. 100 cycles.
N_f = 2500 cycles.

Figure 3 - Scanning electron micrograph of surface of failed specimen.
Total strain range 0.75%. 25°C. Ramp
time 2.5 sec. No hold. 2550 cycles.
N_f = 2500 cycles.

The main difference between the fatigue of solder at high strain ranges (0.75% total strain) and low strain ranges (0.30% total strain) is in the absence of coarse slip lines and transgranular cracks associated with them in the latter. Figure 4 shows some rather faint evidence of slip lines but intergranular cracks were already noted at approximately N/N_f = 0.05. A micrograph (Fig. 5) made of the surface of a failed specimen fatigued at 0.30% total strain range does not show any evidence of transgranular fracture. Thus, the break in the lines in the log fatigue life versus log plastic strain range plots (Fig. 1) can be attributed to the change in the fracture mode from intergranular at low strains to mixed transgranular-intergranular at high strains.

 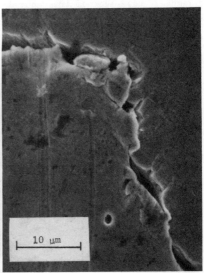

Figure 4 - Scanning electron micro-graph of surface replica. Total strain range 0.30%. 25°C. Ramp time 2.5 sec. No hold. 8000 cycles. N_f = 29000 cycles.

Figure 5 - Scanning electron micro-graph of surface of failed specimen. Total strain range 0.30%. 25°C. Ramp time 2.5 sec. No hold. 30000 cycles. N_f = 29000 cycles.

Specimens which were tested at elevated temperatures were also observed under optical and electron microscopes. As at low temperatures, slip lines and cracks along some slip lines were detected in specimens tested at high strain ranges. The failure mode at low strains remained intergranular at all temperatures.

The results indicate that fatigue life of this solder at 80°C essenti-ally does not differ from the fatigue life at 100°C. The fatigue life at these temperatures is the lowest of all temperatures tested. A gradual de-crease in the fatigue life with increasing temperature was observed as the temperature was increased from 25 to 80°C; however, the fatigue life de-creased when the temperature was decreased from 25 to 5°C. While we are not able to explain with certainty the complex relation between fatigue life and temperature observed in the fatigue fracture of 96.5Pb-3.5Sn solder, the change in fatigue life with increasing temperature may be associated in part with the phase transformation. For this solder, tin completely dissolves in the matrix above approximately 75°C, and above this temperature only one phase is present. The volume fraction of the tin phase increases on cooling below this temperature. Similar to our results, no difference in fatigue life at 120 and 150°C was observed in 95Pb-5Sn solder (3). No microstructural changes were observed when the testing temperature was reduced to 5°C. The small reduction in fatigue life at constant strain range may possibly be due to the low water pressure in the dry air at 5°C. Lead has a higher fatigue life in dry compared to wet air (20).

While practically no change in the number of cycles to failure was observed on varying frequency in the comparatively high frequency region, a reduction in the number of cycles to failure with decreasing frequencies below 10^{-2} Hz was found at different strain ranges at 25, 50 and 80°C. As an example, the effect of frequency on the number of cycles to failure at 25°C and 0.75% total strain is shown in Fig. 6.

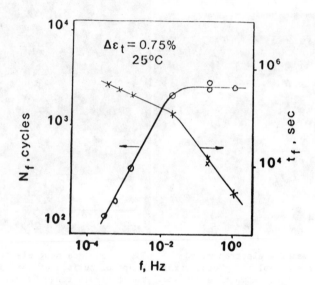

Figure 6 - Effect of frequency on number of cycles to failure and time to failure (t_f).

Figures 7 and 8 show that at low frequencies fatigue failure of this solder at 0.75% total strain is mainly by the intergranular mode although there is some transgranular cracking along slip bands as at higher frequencies.

2. Tests with Hold Time

Figure 9 shows that hold time at maximum strain (tensile stress hold time) in tests with a constant ramp time (t_r) equal to 2.5 seconds has a dramatic effect on the number of cycles to failure. The decrease in N_f is most rapid for small hold times. As in tests with low frequency continuous cycling, the separation of grains along boundaries seems to be the main mode of failure in tests with tensile hold time (Fig. 10).

A limited number of tests with a hold time at zero strain, i.e., with a

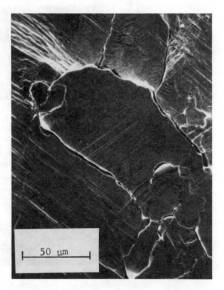

Figure 7 - Scanning electron micrograph of surface replica. Total strain range 0.75%. 25° C. Ramp time 60 sec. No hold. 800 cycles. N_f = 1230 cycles.

Figure 8 - Scanning electron micrograph of surface of failed specimen. Total strain range 0.75%. 25° C. Ramp time 60 sec. No hold. 1300 cycles. N_f = 1230 cycles.

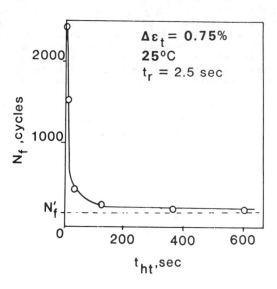

Figure 9 - Effect of tensile hold time on fatigue life.

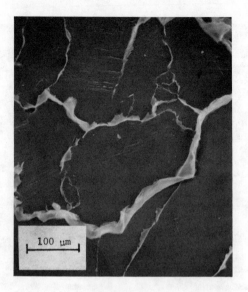

Figure 10 - Scanning electron micrograph
of surface replica. Total strain range
0.75%. 25°C. Ramp time 2.5 sec. Tensile hold time 90 sec. 200 cycles.
N_f = 330 cycles.

compressive stress hold time, and a ramp time of 2.5 seconds, indicated no or
almost no effect of such hold time on fatigue life of the 96.5Pb-3.5Sn solder.
For example, in a test with 30 seconds of compressive hold time, the specimen
survived more than 2500 cycles while N_f values for tests without hold were
equal to 2200-2900 cycles; N_f for test with 30 seconds tensile hold and the
same 2.5 seconds ramp time was equal to 450 cycles. Introduction of a compressive hold time, however, dramatically reduces the number of cycles to
failure when it is combined with a tensile hold time. Figure 11 shows how
variable compressive hold time affects the number of cycles to failure for
solder in tests with constant ramp time of 2.5 seconds and with fixed tensile
hold times of 120 and 360 seconds. Line 1 in this figure corresponds to

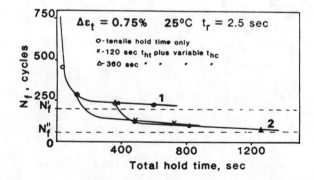

Figure 11 - Effect of total tensile and compressive
hold times per cycle on number of cycles to failure.

134

tests with only tensile hold time, line 2 corresponds to tests with combined tensile and compressive hold times. It is evident that as in tests with just variable tensile hold time, reduction in the number of cycles to failure is most dramatic during the first few minutes of compressive hold application.

The small effect on the fatigue life of 96.5Pb-3.5Sn solder of compressive hold time alone introduced in the cycle with a short ramp time may possibly be due to the expectation that no or almost no grain boundary separation occurs during the fast ramp as well as during the compressive hold.

According to C. H. Wells and C. P. Sullivan (16), voids in Ni-base superalloy nucleate during tensile hold and then grow by vacancy diffusion. When voids become large enough, they are assumed to grow by plastic deformation and to propagate under cyclic loading along grain boundaries. In the case of tensile hold, voids were assumed to be rounded; in compression, the voids were believed to grow into elongated flat cracks. The work required to fracture a grain boundary is lower in the case of the presence of flat crack-type void because stress concentration at the void tip is higher and because such elongated voids cover a higher fraction of grain boundary than rounded voids of the same volume. While this theory concerning the formation of voids in tension and propagation of voids in tension and compression appears to explain the effect of tensile and compressive hold times on the fatigue life of 96.5Pb-3.5Sn solder (formation of voids in this solder during creep was documented by H. J. Frost et al. (17)), a higher rate of void propagation during compressive hold than during tensile hold does not seem to be appreciable in 16.5Pb-3.5Sn solder. At 0.75% total strain at room temperature, the number of cycles to failure in a test with 360 seconds of tensile hold and 120 seconds of compressive hold and in a test of 120 seconds tensile hold and 360 seconds compressive hold was found to be practically the same (130 and 140 cycles respectively).

The saturation in fatigue life of 96.5Pb-3.5Sn solder with increasing tensile or combined tensile or compressive hold times per cycle may be attributed in part to the saturation in the relaxation strain per cycle as done for 304 and 316 stainless steels and for Incolloy 800 at elevated temperature by J. Wareing (18). He assumed that surface cracks initiate and propagate during high rate fatigue plastic strain, $\Delta\dot{\varepsilon}_p$, and the initiation and propagation of grain boundary voids occur during low rate tensile relaxation strain, $\Delta\dot{\varepsilon}_{pr}$.

Figure 12 shows how the relaxation strain per cycle defined as the difference between the width of a hysteresis loop for a test with a given hold time and the width of hysteresis loop for test without hold time depends on tensile hold time in the test with 0.75% total strain at room temperature. The value of relaxation strain per cycle increases very rapidly during the first few minutes; later the increase is much slower and the relaxation strain tends toward a constant value. The effect of hold time on the relaxation strain per cycle and on the number of cycles to failure is similar; therefore, the saturation in fatigue life may be partially related to the saturation in the relaxation strain per cycle. The strain rates associated with continuous cycling in the present research varied from 4×10^{-6} to 3×10^{-2} sec^{-1}, while the highest strain rate associated with the stress relaxation during tensile hold time was approximately equal to 10^{-6} sec^{-1}.

The relaxation strain per cycle as a function of variable compressive hold time when added to the cycle already containing 360 seconds of tensile hold time in a test with 0.75% total strain at room temperature is depicted in Fig. 13. It is evident that compressive hold, in addition to tensile hold time, leads to another saturation in the relaxation strain per cycle (one saturation was observed with only tensile hold time (Fig. 12)). Thus,

135

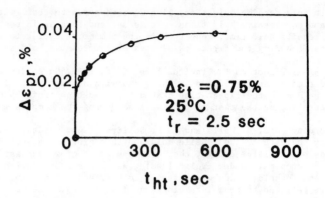

Figure 12 - Relaxation strain per cycle versus tensile hold time.

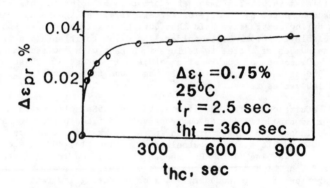

Figure 13 - Relaxation strain per cycle versus compressive hold time.

this saturation in the relaxation strain per cycle may possibly be the origin for the saturation in fatigue life with increasing combined tensile and compressive hold times.

The decrease in the number of cycles to failure and saturation in fatigue life with increasing tensile hold time are affected by environmental attack at the surface of solder. Fatigue tests in vacuum gave a higher fatigue life for 96.5Pb-3.5Sn solder than in air; however, as in air, there was saturation in fatigue life with increasing tensile hold time (19).

<u>Summary</u>

1. There are two mechanisms for 96.5Pb-3.5Sn solder isothermal fatigue failure depending upon experimental conditions. At high frequency the failure mode was found to be intergranular at plastic strains below approximately 0.3% and mixed transgranular-intergranular at the higher strains

Change in the fatigue failure mode leads to two lines in the Coffin-Manson plots.

2. Tensile hold time and combined tensile and compressive hold times were found to reduce dramatically the number of cycles to failure. Decreasing frequencies below 10^{-2} Hz also reduced the number of cycles to failure. Under these conditions, the mode of fracture was predominantly intergranular.

Acknowledgements

The authors wish to thank Mr. L. Lawson for conceiving and building the heating-cooling apparatus used in this research.

Appreciation to the IBM Corporation is expressed for sponsoring this research. Use of the Central Facilities of Northwestern University's Materials Research Center under the NSF-MRL program (Grant Nos. DMR82-16972 and DMR85-20280) is gratefully acknowledged.

References

1. J. H. Lau and D. W. Rice, "Solder Joint Fatigue in Surface Mount Technology", Solid State Tech., 28(10)(1985), 91-104.

2. J. F. Burges et al., "Solder Fatigue Problems in Power Packages", IEEE CHMT-7, 1984, no. 4:405-410.

3. H. S. Rathore, R. C. Yih, and A. R. Edenfeld, "Fatigue Behavior of Solders Used in Flip-Chip Technology", J. Testing Evaluation, 1 (1973) 170-178.

4. W. Hofmann, Lead and Lead Alloys (Springer-Verlag, New York, 1970), 95-100.

5. A. R. Ellozy, P. M. Dixon, and R. N. Wild, "Combined Low-Cycle Fatigue and Stress Relaxation of Some Pb-In Alloys and 96/4PbSn Alloy at Room Temperature", Proceedings of the Second International Conference on Mechanical Behavior of Materials, (1976), 903-907.

6. J. F. Eckel, "Influence of Frequency on the Repeated Bending Life of Acid Lead", Proceedings ASTM, 51 (1951) 721-740.

7. G. R. Gohn and W. C. Ellis,"The Fatigue Tests as Applied to Lead Cable Sheath", Proceedings ASTM, 51 (1951) 721-740.

8. H. D. Solomon, "Fatigue of 60/40 Solder", Proceedings, 35th Electronics Components Conference, IEEE (1986) 622-629.

9. H. J. Shah and J. H. Kelly, "Effect of Dwell Time on Thermal Cycling of the Flip-Chip Joints", ISHM (1970) 3.4.1.-3.4.6.

10. B. N. Agarwala, "Thermal Fatigue Damage in Pb-In Solder Interconnections", 23rd Annual Proceedings, Reliability Physics, IEEE/IRPS (1985) 198-205.

11. J. K. Tien, S. V. Nair, and V. C. Nardone, "Creep-fatigue Interaction in Structural Alloys", Flow and Fracture at Elevated Temperatures, ASM, Metals Park, Ohio (1985), 179-214.

12. Vander Voort, <u>Metallography, Principles and Practice</u> (McGraw-Hill Book Co., New York, N. Y., 1984), 581.

13. L. Lawson, "Thermomechanical Fatigue Testing of 97Pb-3Sn Solder" (M.S. Thesis, Northwestern University, Evanston, IL, June 1986).

14. D. Raynor and R. P. Skelton, <u>The Onset of Cracking and Failure Criteria in High Temperature Fatigue Testing</u> (Elsevier Applied Science Publishers, London and New York, 1985), 143-145.

15. K. Bhanu Sankar Rao et al., "On the Failure Conditions in Strain-Controlled Low Cycle Fatigue", <u>Int. J. Fatigue</u>, 7 (1985) 141-147.

16. C. H. Wells and C. P. Sullivan, "Interaction Between Creep and Fatigue in Udimet 700 at 1400° F", <u>Fatigue at High Temperature</u>, <u>ASTM STP 459</u> (ASTM, 1969), 59-74.

17. H. J. Frost et al., "Flow and Fracture Mapping of Tin-Lead Alloys", (Report to IBM Corp., 1985).

18. J. Wareing, "Creep-Fatigue Interaction in Austenitic Stainless Steel", <u>Metall Trans. A</u>, 8(A)(1977) 711-721.

19. R. Berriche, "Environmental Effects in Fatigue of 96.5Pb-3.5Sn Solder" (M.S. Thesis, Northwestern University, Evanston, IL, 1986).

20. K. U. Snowden, "The Effect of Atmosphere on the Fatigue of Lead", <u>Acta Metall.</u>, 12 (1964) 295-303.

STRESS AND STRAIN CONTROLLED LOW CYCLE FATIGUE OF PB-SN SOLDER

FOR ELECTRONIC PACKAGING APPLICATIONS

B.C. Hendrix and J.K. Tien

Center for Strategic Materials, HKSM
Columbia University, N.Y., N.Y. 10027

S.K. Kang and T. Reiley

IBM, T.J. Watson Research Center
Yorktown Heights, N.Y. 10598.

Abstract

The low cycle fatigue (both stress and strain controlled) properties of
lead-tin alloys have been studied and analyzed with respect to the alloy's
microstructure. Tensile results show a saturation of strain rate effects
above rates of about 10^{-3} sec^{-1}. Stress controlled testing reveals a power
law dependence of creep rate on stress, a much larger mean stress effect than
stress range effect, and a minimum strain rate which decreases as the
frequency increases. Varying waveforms and per cycle data confirm that the
frequency dependence is controlled by an anelastic strain recovery mechanism.
Testing at 0°C reveals a strong sensitivity to testing temperature. R=0
strain controlled testing revealed that the mean stress relaxes toward zero
with a frequency dependence which is controlled by the storage of time dependent strain during both the ramp and the hold at strain. Plastic strain
increments have been imposed with variable ramp rates, i.e. yield stresses.
Metallography characterizes the structure of the alloy.

Introduction

The current drive toward more dense packaging of micro-electronic equipment is leading to new configurations of electrical connections. Some of
these configurations require the solder of the joint to carry the full
mechanical load of the connection. This load is cyclic and high, often producing significant plastic strains in each cycle. The temperatures at which
these joints are used are quite high for eutectic lead-tin, e.g., 0°C is 0.60
homologous (temperature normalized to the melting temperature), 50°C is 0.71
homologous. These conditions define a rather severe creep-fatigue environment.

Studies have been performed which rank solder properties in terms of
tensile or shear strength, creep resistance, and thermal fatigue(1-4), but it
has been pointed out (5,6) that these rankings may vary from test to test and
from bulk to joint configurations. Some of these variations could be a result
of microstructural inconsistencies as joints are cast and not all studies
are, or could be sufficiently well documented microstructurally. In addition, neither can a cast microstructure be adequately reproducible in some
cases. But it is also necessary to determine which mechanical properties best
describe how the solder will behave in actual service.

This investigation has been designed to separate the ambient temperature deformation mechanisms involved in eutectic lead-tin and to correlate these mechanisms to microstructure. Earlier testing in this investigation (7) began characterizing the bulk eutectic mechanical behavior using stress controlled trapezoidal loading waves with varying hold times (frequencies), mean stresses, and stress ranges. The frequency dependence was found to be controlled by anelastic strain storage and recovery. The nonrecoverable strain rate depends on a power law of stress. Other work has found similar frequency dependence in solder joint specimens(8).

The current study extends the variables under investigation to include the effects of unbalanced on- and off-load trapezoidal loading waves and strain controlled testing including large increments of plastic strain. The microstructure is being carefully documented although with cast material it is difficult to be both thorough and accurate as there are significant variations between different specimens as well as within a single specimen.

EXPERIMENTAL PROCEDURE

The eutectic lead-tin alloy used in this study is a commercial grade solder alloy of nominal composition in weight percent of 63Sn-37Pb. Compositional analysis for two lots, C and E, of the alloy are shown in Table 1. The alloy was supplied in rectangular bar form which was then remelted and cast into a 15.9mm diameter cylindrical mold at room temperature. The cyclindrical ingot was then machined to the finished buttonhead specimen with 25.4mm gage length and 6.35mm gage diameter. All specimens were aged at room temperature for at least one month prior to testing.

The metallographic structure of the eutectic alloy specimens was examined to group them according to grain size and grain morphology. Early mechanical tests exhibited some scatter in results because of microstructural variations. Thereafter metallography was performed before testing to minimize microstructural effects on the test results. The best results for microstructural characterization were obtained by electropolishing, at -10°C to 25°C, with a 10% Perchloric-Ethanol solution, at 12-40 volts for 5-30 seconds.

The specimens were then characterized by both optical microscopy and scanning electron microscopy(SEM). Characterization includes grain size, grain shape, eutectic morphology, and the existence of proeutectic tin or lead. An example of various eutectic morphologies existing in close proximity along with an example of proeutectic tin is shown in Figure 1.

Although the main focus of the study was on the creep-fatigue interaction in these lead-tin alloys, tensile tests were performed on each alloy at various strain rates. These were necessary in order to insure that the maximum stress imposed on the stress controlled tests was below the 0.2% offset yield stress at the rate of loading for the particular waveform and frequency being used. In some of the strain controlled tests the yield strength was exceeded by definition. It was also necessary to find a yield strength plateau in order to separate the thermal from the athermal plastic flow.

TABLE 1.   EUTECTIC ALLOY COMPOSITION
(all values in weight percent)

Alloy	Sn	Pb	Cu	Bi	In	Sb	Ag	Al
C	63.1	Bal.	0.0001	0.0002	0.0001	-----	0.0001	-----
E1	62.0	Bal.	-----	-----	-----	-----	-----	-----
E8	61.9	Bal.	-----	-----	-----	-----	-----	-----
E9	61.2	Bal.	-----	-----	-----	-----	-----	-----

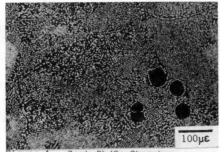

Figure 1. Cast Pb/Sn Structure.

All of the mechanical testing was performed on a computer controlled servo-hydrolic testing machine. The strain measuring system is capable of resolving strains of 15με (1με = 10^{-6} mm/mm). The standard trapezoidal wave form used in this study had a ramp rate which was 0.25 times the hold time as shown in Figure 2. The variables of mean stress and alternating stress were more thoroughly investigated than previously with mean stresses from 1MPa to 21MPa and stress ranges from 13MPa to 28MPa at 25°C and a frequency of 0.167Hz. Limited testing was performed with only on-load hold times or only off-load hold times in the trapezoidal waveforms (Fig. 2) to separate the increase in life from anelastic relaxation from the possibility of cyclic strengthening. Tests were performed with fast ramping trapezoidal waveforms to separate the anelastic from the nonrecoverable strain. The standard trapezoidal waveform was used for strain controlled testing where all tests were controlled to R=0 conditions at 25°C.   Strain ranges from 2000με to 20,000με and frequencies from 0.00167Hz to 1.67Hz were examined.

Results and Discussion

Stress Controlled Testing

The results of varying mean stress on the life are shown in Figure 3. Although these results show significant scatter because of the microstructural variations in the cast structures, it is clear that life is more dependent on mean stress than on stress range. Where a factor of 2 in stress range can make a difference of perhaps a factor of 5 in life, a factor of 2 in mean stress produces a change of around 2 orders of magnitude in life.

This indicates that most of the damage is occurring from thermal creep type mechanisms rather than localized strain and crack propagation mechanisms at this temperature, 25°C, and frequency, 0.167Hz. Previous higher frequency tests (up to 1.67Hz) on this alloy (7) exhibited more fatigue-like behavior and so it is expected that more stress range dependence would be exhibited at higher frequencies.

Envelope strain is plotted against the time spent on load in Figure 4 for three specimens which had been tested using hold time on-load, balanced, and hold time off-load trapezoidal waveforms. All had the same hold time and ramp rate. The time on load, which is defined here as the time spent above the mean stress, is used rather than time or number of cycles. This allows the comparison

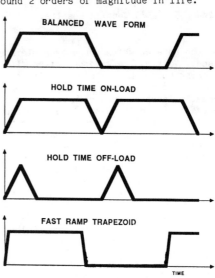

Figure 2. Wave Form Controls.

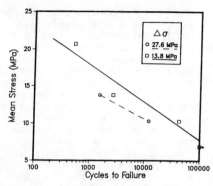

Figure 3. Life for Mean Stress at 0.167Hz and 25°C.

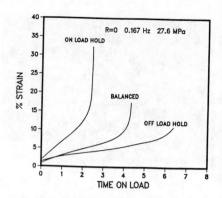

Figure 4. Envelope Strain for Various Wave Forms.

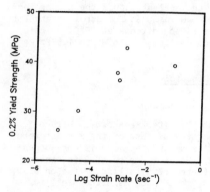

Figure 5. 25°C Tensile Results.

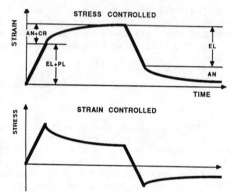

Figure 6. Storage and Recovery of Strain.

of the rate at which nonrecoverable strain is stored for each waveform being used. Varying the waveform gives different relative amounts of time spent storing total thermal strain (time on-load) as compared to the time spent recovering the anelastic component of the strain (time off-load). There is a drastic increase in strain rate with relative time spent on load, which indicates that the increased life with increased frequency is a result of the anelastic strain storage preventing the storage of nonrecoverable strain. In the case where no time is spent recovering the anelastic strain during unloading, the anelastic mechanism is unable to prevent the storage of nonrecoverable strain, and so nonrecoverable strain is being stored right from the beginning of the on-load hold. At the extreme of the bottom-heavy waveform, there is always sufficient time to recover anelastic strain during the off-load period, but during the very short on-load time the storage of anelastic strain prohibits the storage of much nonrecoverable strain. (This is a phenomenological description of the various components of the strain and their interaction. A mechanistic explanation cannot be offered yet although this type of behavior has been explained mechanistically for other systems (9).)

The tensile tests shown in Figure 5 demonstrate that the yield stress is independent of strain rate above about 10^{-3} sec^{-1}. This indicates that essentially no time dependent plastic flow occurs at these ramp rates and so these ramp rates can be used to separate the thermal from the athermal plas-

142

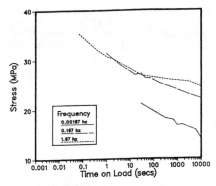

Figure 7. Relaxation in 2000µε Control Test.

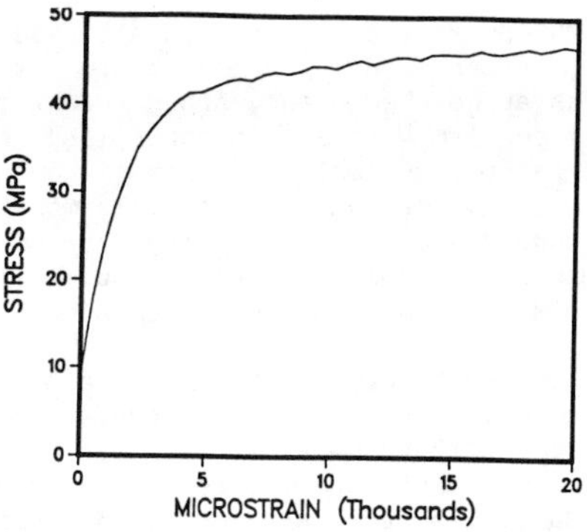

Figure 8. Tensile Results at 0.1 sec^{-1}.

tic strain. This was used as the basis for the fast trapezoidal wave shown in Figure 2 which was used for further stress controlled testing.

The strain storage and recovery during the steady state stage of the creep tests was analyzed using data from individual cycles as shown in Figure 6. The average values of measurements from 11 cycles are shown in Table 2 from a 27°C test and a -1°C test. Anelastic strength is usually defined as the ratio of the anelastic strain stored in an infinite amount of time at a certain stress to the amount of elastic strain stored at that same stress. In this case we refer to a frequency dependent strength which is the ratio of the anelastic strain recovered in a cycle to the elastic strain recovered in a cycle. It should be noted that the decrease with temperature of the elas= tic strain stored at 27.6 MPa is probably a result of the phase change at about 13°C as well as modulus variations.

Table 2. Temperature Dependence of Strain

Apparent Temperature (K)	272	300	Activation
(°C)	-1	27	Energy
Time Independent Strain	1058µε	1558µε	
Total Stored Strain	124µε	279µε	
Recovered Strain	132µε	279µε	
Anelastic Strength	0.125	0.179	0.11eV
Net Stored Strain	<15µε	254µε	
Minimum Strain Rate	0.985µε/sec	77.8µε/sec	0.95eV
Lead Self Diffusion			1.11eV (10)
Tin Self-diffusion			1.10eV (10)

The apparent activation energy taken from these 2 data points implies that for a 2°C variation in test temperature around room temperature, the rate of nonrecoverable strain storage will change by about 10%. This empha= sizes the need for careful control of test conditions. It can also be seen that the activation of the anelastic mechanism requires much less energy than the nonrecoverable strain storage.

Strain Controlled Testing

All strain controlled testing was conducted from 0 to a maximum strain, R=0. Stress relaxation curves are shown in Figure 7 for various frequencies

143

and 2000µε maximum strain. In all cases, although the minimum stress started at 0, it became negative as strain was stored. This continued until the mean stress approached 0 and, after cracks became significant, even dropped below 0. A problem with strain controlled testing is in defining an easily measurable failure criterion or rate of damage accumulation. For this study, failure was admitted when the tensile modulus was severely decreased as a result of reduced cross section. The rate of stress relaxation measures the storage of unrecovered plastic strain, but it is unclear as to whether strain storage of this sort is damaging in terms of the life of the specimen.

The most notable effect of raising the test frequency is the increase in the average stress in a half cycle. This is a natural consequence of allowing more time to relax in each cycle. At the higher frequencies, the time dependent strain that is being stored during the hold should not be storing any damage as it is all recovered during the hold of opposite sign. But the higher frequencies also have higher ramp rates which disallow the storage of time dependent strain during the ramp and thus cause some greater amount of athermal plastic flow to occur during the ramp. The storage of athermal plastic strain occurs at stresses much lower than the 0.2% offset as can be seen in Figure 8.

Larger strain ranges store a significant amount of plastic strain on the ramp up so that a mean stresses of 0 is obtained after the first cycle. In these cases both athermal and thermal components of the nonrecoverable strain contribute to the stress relaxation. The life results of these tests are shown in Table 3. The frequency dependence at these higher strain tests has not been examined yet, but it should reflect the ductility available from either of the two nonrecoverable strain components that were identified in the stress controlled testing.

Table 3.   Strain Controlled Testing at 0.167 hz

Strain Range	Cycles to Failure
2,000µε	>30,000
4,000µε	12,963
20,000µε	185

Summary

The low cycle fatigue behavior of Pb-Sn solder is controlled by the interaction of at least two thermal and two athermal strain storage components. These are nonrecoverable creep strain, anelastic strain (recoverable), elastic strain, and athermal plastic strain. Time dependent components are not observed at strain rates above about 10^{-3} sec^{-1}. The storage of anelastic strain prevents the storage of nonrecoverable strain which leads to the suppression of creep at higher test frequencies. Nonrecoverable creep follows a power law with stress. Temperature variations of even 2°C can significantly affect test results. Strain controlled tests relax to a zero mean stress with a frequency dependence dominated by the storage of time dependent strain. In strain controlled tests the athermal plastic strain can be a significant contributor to damage. Even though some packaging applications apply conditions closer to strain controlled testing than stress controlled testing, stress controlled testing yields more specific information about the specific mechanisms storing the strain. The microstructural variations of a cast structure are significant to the fatigue of this alloy.

Further testing needs to address the following:
1. The total anelastic strain at saturation, i.e., the accepted anelas-

144

tic strength and further characterization of the anelastic strain in terms of various stresses and more temperatures.

    2. The nonrecoverable strain storage at more temperatures and stresses at lower frequencies to separate it from the anelastic strain.

    3. The effects of frequency on tests with large increments of plastic strain.

    4. The case below 13°C.

Acknowledgement

We thank the IBM T.J. Watson Research Laboratory for sponsoring this research. We acknowledge Mr. Erik Schwartzkopf for his tireless problem‑solving.

REFERENCES

1. D.O. Ross, in Proceedings of 4th annual International Electronics Packag‑ing Conference, Baltimore, October 1984, pp. 181‑187.

2. R. Horiuchi, A.B. El‑Sebai, and M. Otsuka, Phys. Stat. Sol., 21 (1974) K89.

3. R.N. Wild, "Properties of Some Low Melt Fusible Solder Alloys," presented at INTERNEPCON, Brighton, England, October 1971.

4. D. Grivas, K.L. Murty, and J.W Morris, Jr., Acta Met., 28 (1979) 731.

5. F.A. Mohamed and T.G. Landgon, Acta Met., 23 (1975) 697.

6. R.N. Wild, Welding Research Supplement, November 1972, pp. 521s-526s.

7. R.C. Weinbel, J.K. Tien, R.A. Pollak, and S.K. Kang, J. Mat. Sci., to be published.

8. R.C. Weinbel, E.A. Schwartzkopf and J.K. Tien, Scripta Met., to be published.

9. V.C. Nardone and J.K. Tien, Met. Trans. A, 17A (1986) 1577.

10. J. Askill, Tracer Diffusion Data for Metals, Alloys, and Simple Oxides (New York, N.Y.: Plenum Publishing Corporation, 1970), 35,39.

LOAD SEQUENCE EFFECTS ON THE DEFORMATION OF ISOLATED

MICROPLASTIC GRAINS

M.R. James and W.L. Morris

Rockwell International Science Center
1049 Camino Dos Rios
Thousand Oaks, CA   91360

Abstract

Strains measured in individual large grains of Al 2219-T851 are used to deduce the local constitutive behavior that controls plastic flow under spectrum loads. A composite model for the deforming grain is most consistent with experiment. The material apparently contains two intermingled dislocation structures whose flow is controlled by different yield criteria. Special load sequences are used to determine the yield criterion best descriptive of each component, based on strains measured over 100 μm gauge lengths in individual grains.

Introduction

Even for "elastic" loading, the role of inhomogeneous localized plasticity in fatigue has been demonstrated repeatedly (1-3). Recent theories of inhomogeneous plasticity in tensile and cyclic loading principally consider large strains and show the importance to plastic flow of inhomogeneity in alloy microstructure (4,5) (e.g., granularity) and of local instabilities in deformation (6). When dislocation phenomena have been used to model the constitutive response of local elements in a material, it is often supposed that the underlying events of dislocation generation, saturation and recovery are sequential and conditioned on the local total strain range. However, for the small plastic strains common to most structural fatigue failures, the dislocation structures themselves tend to be inhomogeneous. Dislocation channels or bands penetrate a matrix of dislocation tangles, creating a microscopic composite (7-9). The local stress state in such subelements, their local mechanical properties and constitutive response may all differ, and each quasi-homogeneous material element will "remember" prior deformation and react to factors beyond total local strain and stress. A complex interaction between load sequence, local stress-strain response and rates of fatigue crack initiation can be anticipated for all such materials.

Recent studies of the deformation of individual large grains in an overaged aluminum alloy (10-14) suggest the importance of a "composite" dislocation structure in modeling the local strain - external stress response. The flow of each isolated microplastic grain is constrained by the nearly elastic surface, creating reaction stresses reminiscent of those at the root of a locally plastic notch. Procedures have been developed (10,11) to determine the mechanical properties of isolated plastic grains from strains measured over microscopic ($<$ 100 μm) gauge lengths. The local internal stresses that might affect the measured property values are handled by load shedding, which either reduces the stresses (12), or allows the resulting plastic hysteresis to be analyzed (10,11). After fatigue of Al 2219-T851, marked hardening of the plastic flow is observed below a critical stress, above which there is simultaneous and dramatic strain softening (12). Does this softening herald the development of banded slip, seen only by high resolution strain field mapping (13) in 2219, or is it better explained by the saturation and recovery mechanism of a single subcomponent of the material, utilized by Estrin and Kubin (6) in modeling inhomogeneous plasticity?

To resolve this question we examine the local constitutive behavior of individual grains by studying their response to several load sequences. In particular, we determine what constitutive behavior, in terms of motion of the yield locus, must be assigned to each element of a two-component composite in order to achieve the observed stress-strain reaction to mean stress and tensile reloadings found experimentally.

Background

When Al 2219-T851 is loaded cyclically below its 360 MPa cyclic yield strength, the initial strains in a 300 μm grain found over a tensile increment are small, but discernible above the experimental measurement error, at stresses greater than 70 MPa (σ_ℓ) (Fig. 1a). A few hundred cycles at ± 270 MPa changes the local properties of the grain, creating an upper flow stress (σ_u) near 200 MPa, strain hardening below σ_u, and softening above (Fig. 2a) (12). The technique used to measure strains over the 100 μm gauge lengths required for this study (see Refs. 10, 11, 15, 16) employs displacements measured between two points of zero external load in a loading sequence. The total strains calculated from the displacements reflect reactions to local plasticity (residual), rather than the elastic strains stemming from an external load. If the residual strains are less than ~ 4 × 10^{-5} the reaction stresses will typically be sufficiently small that data such as in Fig. 2a will depict the true local stress-strain response. If the residual strains are large, deformation models are needed to separate the elastic reaction strain and plastic strain components, in order to identify flow points and calculate strain hardening coefficients.

Deformation Models

Other studies on Al 2219-T851 (10,11) indicate that in the overaged state, local mechanical properties do not vary greatly across a grain so that an assumption of uniform plastic deformation is adequate for fully reversed loading. Two limiting cases of the flow surface locus have been examined for microplastic grains, namely: 1) a "stationary" surface (S) independent of prior load; and 2) a "kinematic" surface (K), which moves freely with the maximum stress excursion on the previous load reversal (11). With the S model, the grain has no memory of being deformed other than its instantaneous plastic strain and the resulting reaction to its constraint by the grain boundaries, while with the K model a maximum possible "back stress" is remembered in addition to plastic strain and boundary constraints. The deformation of microplastic grains for which σ_u and σ_ℓ are both stationary or both kinematic has been considered elsewhere (11). The presence of a K modelled behavior for σ_u has been ruled out as giving an unacceptable description of strain transients following a sudden change in cyclic stress (11), but the possibility of a K model of σ_ℓ and an S model of σ_u has not been examined.

For this study we model the latter situation by defining a composite material for the grain, assumed to experience the same local stress σ^ℓ in each element. Each of the two components (i = 1,2) is assigned a "plastic" modulus for plastic strain above its respective flow stress, giving a plastic strain increment

$$\Delta \varepsilon_p(i) = \frac{\Delta\sigma^\ell}{E_p(i)} .$$ (1)

The strains in the two elements are summed; thus E_p contains a hidden dependence on both the strain hardening and the relative volume of the two components. The flow criteria are applied independently to each element in a numerical calculation (based on equations in Ref. 10) of the strains and stresses at each load reversal. This allows σ_ℓ to be defined according to a K model and σ_u to an S model status. (The complete formulation of this model will be discussed elsewhere.)

In addition we have considered a model describing a nonuniformly deforming grain represented by an interior enveloped by a shell of different mechanical properties near the grain boundary. The stress-strain behavior of such dual-domain systems has been described by Morris et al. (10). The constitutive equations for a dual-domain grain were implemented numerically for an arbitrary load sequence. In the following we use U and D to denote uniform and dual-domain grain models; the entry K/S refers to a K model of σ_ℓ and an S model of σ_u. For example, D-S/S is a dual-domain model with a stationary flow surface locus for both σ_ℓ and σ_u.

Results

Deformation in a 300 μm grain before σ_u develops during fatigue is examined in Fig. 1. The sample was cycled at ±270 MPa for 100 cycles and then the load was dropped to σ_a as shown in the insert of Fig. 1a. Measurement of W_o was made after 5 cycles at σ_a to allow for equilibrium of the local stress-strain response. Repetition of load shedding for multiple values of σ_a has been found most useful to determine the local mechanical properties of the grain (11). Material values were chosen for U-S/S and U-K/S models to fit the observed residual strains in Fig. 1a. However, for these same parameters, only the U-K/S model depicts the strain behavior for a sequence of tensile reloads to successively higher tensile maxima (σ_{max}) (Fig. 1b). The large increase in the σ_ℓ flow locus that occurs with a K model is necessary to prevent significant grain deformation on reloading.

After a few hundred fatigue cycles, σ_u appears (Fig. 2a). Residual strains measured during load shedding for fully reversed loading are again described by selection of appropriate material parameters for the two regions of flow for the U-K/S and U-S/S models. Using these coefficients, only the U-K/S model correctly represents the increasing strain in response to a series of stress relaxation cycles around an increasingly more positive mean stress (σ_{mean}) (Fig. 2b). A kinematic increase of the σ_ℓ flow locus suppresses the large internal strains that would develop with increasing mean stress. Another way this result might occur is if flow at the grain boundary lowered the maximum stress experienced by the grain interior. This possibility was examined using the dual-domain model (10). With appropriate selection of material properties for the boundary, the D-S/S model can almost mimic the behavior in Fig. 2b but, for the same material parameters, the representation of the behavior in fully reversed loading (Fig. 2a) is poor.

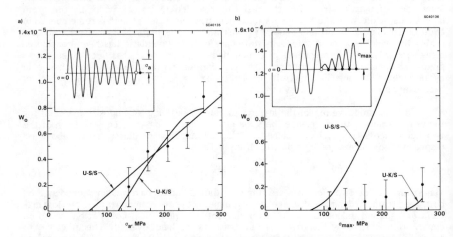

Fig. 1   Local strain behavior in a 300 μm grain after 100 cycles at ± 270 MPa. W_o is a residual strain measured at zero external load in the load sequence (solid circle), relative to an earlier point (open circle). a) As a function of the applied cyclic stress amplitude; b) for a series of loads of increasing tensile maximum σ_{max}, following unloading from 270 MPa.

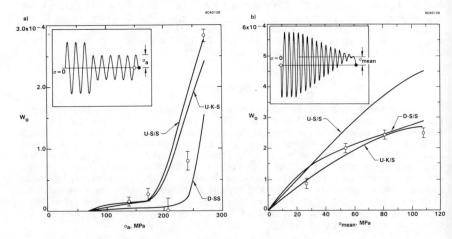

Fig. 2   Local strain behavior in a 300 μm grain after 250 cycles at ± 270 MPa. W_o is a residual strain measured at zero external load (solid circle) relative to an earlier point (open circle). a) As a function of the applied cyclic stress amplitude; b) following a stress envelope converging to a positive mean stress σ_{mean}, relative to an envelope collapsed to a zero mean.

Discussion

In a grain of monolithic properties, if σ_ℓ is less than σ_u for any loading sequence, then the relationship must hold for all sequences. This is simply because each region of flow in a monolithic material will be conditioned by particular levels of strain. To ensure this behavior in a two-stage yielding model, both the lower and upper flow stresses would have to be either stationary or kinematic. In contrast, the strain behavior of single grains we report for the mean stress and tensile reloading experiments is represented well only if, (1) the high strain hardening deformation found just above σ_ℓ occurs at a stress less than σ_u in fully reversed loading, and (2) if σ_ℓ is greater than σ_u after single large tensile excursions. It appears that for this to happen, σ_ℓ must be kinematic, σ_u stationary, and, further, that the grain must consist of a microscopic composite of material elements of two types, one having a flow stress σ_ℓ, the other σ_u. It also suffices to treat the deformation as uniform throughout the grain, so the individual composite subcomponents must be small compared to a 100 μm gauge length. Since a stationary flow locus is tantamount to a large separation between dislocation pinning points, is σ_u then associated with banded slip? This would mean that onset of flow in the bands can require higher stresses than flow in the matrix (above σ_ℓ), an uncomfortable proposition except that flow in the bands involves much larger strains than in the matrix.

For Al 2219-T851 it is becoming clear that the relationship of σ_u to σ_ℓ in individual grains is extremely dependent on the prior load. It is, therefore, unavoidable that the local plastic strains in this material will be highly sensitive to the details of load sequence as well as to the amplitude range. With the assistance of a simple model depicting the way the material properties in the grain evolve with cumulative plasticity (14), it should now be possible to calculate how load sequence will effect the local strains, and hence the progress of fatigue crack initiation in isolated microplastic grains.

Summary

The complexities of the deformation of a large microplastic grain in Al 2219-T851 are revealed by local strain measurements made for external loading other than fully reversed. Two stages of flow are recognized, and a comparison with models of the deformation process suggests the likelihood that each takes place in a distinct subcomponent of the grain. Many of these elements must be intermingled to form a microscopic composite, as the deformation appears to be quasi-uniform over gauge lengths comparable to the grain size. One subcomponent, known to strain harden quickly with fatigue, is best described by a kinematic movement of the yield surface locus with prior loading, suggesting a reaction to short range back stresses in the grain. Deformation of the material is therefore extremely dependent on the prior load sequence. The second component dramatically strain softens with fatigue, and is best described by a "stationary" yield surface locus - suggestive of a long-ranged constraint of flow by grain boundaries. A physical representation of our model equations might be a matrix containing slip bands, although we know for the 2219 aluminum alloy that traditional persistent slip bands are not present. Nevertheless, the single uniform deformation model which incorporates both of these kinds of behavior provides an adequate description of the deformation of the isolated nonplastic grains in the 2219 aluminum alloy that are responsible for fatigue crack initiation.

Acknowledgement

This work was supported by the Naval Air Development Center under Contract No. N62269-86-C-0261. The experimental help of R.V. Inman at the Science Center is gratefully acknowledged.

151

References

1. A. Esin and W.J.D. Jones, "A Statistical Approach to Microplastic Strain in Metal," J. Strain Analysis 1 (1966), 415-421.

2. J. Polak and M. Klesnil, "The Hysteresis Loop, 1. A Statistical Theory," Fat. Engng. Mater. and Struc. 5 (1982) 19-32.

3. M.R. James and W.L. Morris, "The Role of Microplastic Deformation in Fatigue Crack Initiation," Fatigue Mechanisms: Advances in Quantitative Measurement of Physical Damage, ASTM STP-811, Am. Soc. Testing and Materials, Philadelphia, PA, 46-70 (1983).

4. U.F. Kocks, "Kinetics of Nonuniform Deformation," Progress in Matls. Sci. Chalmers Anniversary Issue, ed. J.W. Christian et al., Pergamon Press, Oxford (1981), 185-241.

5. R.E. Stoltz and R.M. Pelloux, "The Bauschinger Effect, Monotoric and Cyclic Hardening in Precipitation Stengthened Aluminum Alloys," Work Hardening in Tension and Fatigue, ed. A.W. Thompson, AIME (1977) 224-229.

6. Y. Estrin and L.P. Kubin, "Local Strain Hardening and Nonuniformity of Plastic Deformation," Acta. Metall. 34 (1986) 2455-2464.

7. H. Mughrabi, "Cyclic Plasticity of Matrix and Persistent Slip Bands in Fatigued Metals, Continuum Models of Discrete Systems, 4," Eds. O. Brulin and R.K.T. Hsieh, North-Holland Publ. Co. (1981) 241-157.

8. H. Mughrabi, "Dislocation Wall and Cell Structures and Long-Range Internal Stresses in Deformed Metal Crystals," Acta Metall. 31 (1983) 1367-1379.

9. H. Mughrabi, "On the Flow Stress of Deformed Metal Crystals Containing Heterogeneous Dislocation Distributions," S. Afr. J. Phys. 9 (1986) 62-68.

10. W.L. Morris, B.N. Cox and M.R. James, "Microplastic Surface Deformation of Al 2219-T851," Acta Metall. 35, (1986) 1055-1065.

11. B.N. Cox, W.L. Morris and M.R. James, "Two-Stage Microplastic Surface Deformation of Al 2219-T851," Acta Metall. 35 (1987), 1289-1300.

12. W.L. Morris, M.R. James and B.N. Cox, "The Evolution of Local Mechanical Properties of Al 2219-T851 During Fatigue," in Fatigue 87, 3rd Int. Conf. on Fatigue and Fatigue Thresholds, June 28-July 3, 1987, Charlottesville, VA, in press.

13. W.L. Morris and M.R. James, "Fundamental Characterization of Surface Microplasticity," Tech. Report, Contract No. DMR-8310652, NSF (Dec. 1984).

14. W.L. Morris, M.R. James and B.N. Cox, "A Phenomenological Model of the Mechanical Properties of Microplastic Grains During Fatigue," Maters. Sci. Engng. (1987), in press.

15. W.L. Morris, R.V. Inman and M.R. James, "Measurement of Fatigue Induced Surface Plasticity," J. Maters. Sci. 17 (1982) 1413-1419.

16. M.R. James and W.L. Morris, "The Fracture of Constituent Particles During Fatigue," Maters. Sci. Engng. 56 (1982) 63-71.

THERMAL FATIGUE OF STAINLESS STEEL *

W. B. Jones, R. J. Bourcier, and J. A. Van Den Avyle

Sandia National Laboratories
Albuquerque, NM 87185

Abstract

Two austenitic steels, 316 Stainless Steel and Alloy 800, have been examined under conditions of both isothermal low cycle fatigue (LCF) and thermomechanical fatigue (TMF). The TMF tests were conducted between 649 and 360°C with a carefully controlled triangular waveform. The LCF tests were performed at 649°C and both kinds of tests were subjected to a strain range of 0.5%. TMF shortened life to 40% for 316 Stainless Steel and to 5% for Alloy 800. The microstructural evolution occurring in both alloys has been examined and we conclude these do not play a role in the life shortening caused by TMF. The TMF does produce asymmetric hysteresis loops with large tensile peak stresses in tests where the maximum temperature corresponded with the peak compressive stress. The influence of TMF on fatigue crack growth rates has been measured and it was found that TMF accelerated crack growth in Alloy 800 and slowed it down slightly in 316 Stainless Steel. The dominant influence of TMF appears to be in fatigue crack initiation, with the tensile peak stress development driving early crack initiation.

* This work performed at Sandia National Laboratories supported by the U. S. Department of Energy under contract number DE-AC04-76DP00789.

Introduction

Thermomechanically induced fatigue is an important part in the service conditions present in structures used at elevated temperatures. This includes tube and piping components of stainless steel used in such energy conversion technologies as nuclear reactors, fossil fueled power plants, and solar central receiver plants. In the first two applications, the number of thermal cycles is purposely kept small to minimize the damaging effects of the thermomechanical cycles; in the third case the diurnal solar cycles and cloud passage make this impossible. The current baseline service condition established for central receiver design is 10,000 cycles accumulation in 30 years plant lifetime. Thermomechanical fatigue (TMF) effects are present in all of these applications but contribute differently in each one depending on how much cycling occurs during service.

Given the difficulty of performing thermomechanical cycling tests, it is not surprising that few thorough studies have been conducted. Some testing has involved moving samples between baths or furnaces maintained at two different temperatures or by using a programmable controller to drive a furnace. Such testing is limited in that direct comparisons to the isothermal low cycle fatigue data base are difficult. References 1, 2, and 3 describe several typical earlier studies.

Both 316 Stainless Steel and Alloy 800 are austenitic alloys commonly used in elevated temperature applications. The 316 Stainless Steel contains 13% Ni and 17% Cr with 2% Mo for added high temperature strength; Alloy 800 is an iron based alloy containing 33% Ni and 21% Cr and was developed to have good corrosion resistance. The higher Cr content also makes this alloy more resistant to sensitization than 316 Stainless Steel making Alloy 800 desirable in welded structures. The strength, fatigue and creep properties of Alloy 800 are at least comparable to the 300 series stainless steels properties (4), leading to use of Alloy 800 as a tubing and piping alloy in nuclear reactors and solar receivers as an alternative to the 300 series stainless steels. In fact, the low strain range isothermal low cycle fatigue (LCF) lifetime data for Alloy 800 are significantly longer than for 316 Stainless Steel (5, 6). Metallurgical studies of Alloy 800 (7-10) have shown that the microstructural evolution of this alloy is more complex and is less well understood than is 316 Stainless Steel (11, 12). This complexity comes primarily from the presence of $A\ell$ and Ti in Alloy 800. These elements may be present (by ASTM specification) at up to 0.6% and can give rise to significant amounts of TiC carbides and γ' ($Ni_3(A\ell,Ti)$) precipitates. There is a competition for the carbon (<0.10% specified) between the two stable carbides, $M_{23}C_6$ and TiC and the extent of TiC formation determines how much ($A\ell$ + Ti) would be available for subsequent γ' precipitation. In addition, final volume fractions of these phases have been shown to be history sensitive (13, 8, 9, 10) with variations in both chemistry and processing history influencing the microstructural evolution and mechanical properties during elevated temperature testing or service. Austenite decomposition in 316 Stainless Steel has been studied and found to constitute only $M_{23}C_6$ carbide precipitation for up to about 103 hours of exposure at 650°C (11, 12).

The ASME Boiler and Pressure Vessel Code and Code Case N-47 which are used to design the stainless steel piping components for these applications do not explicitly treat TMF effects. However, the current rules in N-47 incorporate an important assumption of material behavior: that isothermal cycling at the peak temperature is a conservative estimate of fatigue lifetime under thermomechanical cycling conditions. This is incorporated through the use of strain controlled LCF data to determine the baseline fatigue curves included in N-47. The correctness of this assumption is not

154

critical to the design of fossil fueled plants or nuclear reactors since the alloy performance there reflects predominantly the creep or cyclic creep behaviors. For the solar application, the thermal cycling aspects of material response must be treated correctly. The appropriate way to treat this effect in structural design is not the topic of this paper. Rather, the purpose of this study was to carefully test this assumption and examine the strengthening and fatigue crack growth mechanisms present during TMF.

Experimental Procedure

Starting Materials

Table 1 shows the chemistries of the heats of 316 Stainless Steel and Alloy 800 used in this study; both are fully within the ASTM specifications for those alloys. In addition to the different chromium and nickel contents between the two alloys, several other differences are significant. The 316 Stainless Steel contains about 2% Mo as a solid solution strengthener at elevated temperature which Alloy 800 does not have. The Alloy 800 contains 0.4% each of Ti and Aℓ which are not present in the 316 Stainless Steel.

Table 1 - Alloy Compositions, Weight Percent

	Fe	Cr	Ni	Mo	Mn	C	Ti	Aℓ	Si
316 Stainless(1)	bal	17.2	13.5	2.34	1.86	.057	.02	--	.58
Alloy 800 (2)	46.5	20.0	31.0	--	--	.04	.40	.40	.24

(1) Oak Ridge National Laboratory Reference Heat No. 8092297
(2) Produced by Huntington Alloy Products Division of International Nickel Corp. (Heat HH 8989A)

Optical metallography of the two alloys is shown in Figure 1. The primary difference seen at this scale is grain size: 5 to 15 μm for the Alloy 800, and about 75 μm for the 316 Stainless Steel. The 316 Stainless Steel starting structure reflects the solution anneal treatment (about 1 hour at 1180°C) given this alloy. Transmission electron microscopy (TEM) has shown that essentially all of the carbides have been dissolved and the dislocation density was very low. The Alloy 800 material shows extensive $M_{23}C_6$ carbide precipitation in both optical microscopy and TEM. An earlier study (9) of this heat of Alloy 800 describes the starting condition in more detail. These particles lie along current grain boundaries as well as in arrays which are suggestive of precipitation along prior grain boundaries. A fine distribution (10 to 100 nm particles) of TiC carbides was found in the form of linear arrays which probably mark the location of prior twin boundaries. This alloy received the standard final mill anneal treatment of one hour at 1000°C. This results in recrystallization of the austenitic matrix without dissolution of most of the carbide particles already present.

Mechanical Testing

The tests performed in this study were done on a 245 kN MTS servo-hydraulic test system. The basic test configuration is shown in Figure 2. The test specimen used in this study is axisymmetric (gage 6.35 mm dia. X 12.7 mm length) with button grip ends. Water-cooled split collet grips are

155

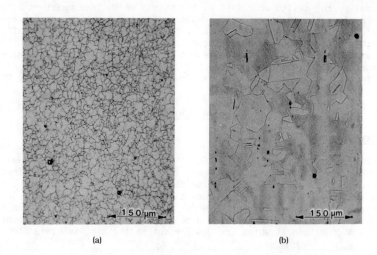

(a)   (b)

Figure 1. Optical microstructures of (a) Alloy 800
and (b) 316 Stainless Steel.

used to provide good axial alignment during fully reversed loading.
Specimen heating is accomplished with a 7.5 kW Lepel RF generator using a
water cooled copper coil to give a temperature gradient of less than 5°C
over the specimen gage length. Temperature is monitored using either K-type
thermocouples or a Vanzetti infrared optical pyrometry system. Strain is
measured using a water cooled MTS quartz rod axial extensometer. For tests
utilizing the optical pyrometry system, the specimen is painted with a flat
black coating (Pyromark) which allows efficient radiation of infrared energy
from the specimen surface (emissivity = 0.93).

    LCF tests were performed using an MTS 410 digital function generator to
provide a triangular waveform for total strain control. The TMF tests were
carried out using a computer controlled test system described in detail
elsewhere (14). Briefly, the system uses a DEC PDP 11/34 computer for data
acquisition and test control through an MTS 433 interface to an MTS 442
controller. Specimen temperature is monitored with the optical pyrometry
system and controlled by varying power to the RF generator through a Leeds &
Northrup Electromax V digital temperature controller and/or activating a
forced air cooling system. All programs used for test control and data
acquisition are written in MTS MultiUser Basic. This computer control allows
simultaneous ramping of strain and temperature, which enables both in-phase
and out-of-phase TMF testing to be accurately performed. The triangular
thermal cycle followed the control signal very closely with no resolvable
overshoot at either temperature peak.

    LCF tests were performed at a fully reversed total strain range of 0.5%
at 649°C for cyclic periods of 10 and 60 seconds. In the interest of
clarity, we will introduce the following nomenclature to define
thermomechanical cycling: testing in which the compressive stress peak
coincides with the maximum temperature will be designated as HTC (High-
Temperature-Compression) while tests in which the tensile stress peak
coincides with the maximum temperature will be designated as HTT (High-
Temperature-Tension). HTC cycling is simply performed by maintaining a
fixed specimen gage length and cycling temperature between the upper and

Figure 2. Testing configuration.

lower limits selected to yield the desired level of thermal expansion
strain. HTT cycling, on the other hand, requires that the specimen gage be
extended and compressed in phase by twice the amount which accompanies
simple free thermal expansion. A temperature range of 360 to 649°C was used
in the TMF tests, which were performed at 0.5% strain range for cyclic
periods of 60 seconds. HTC TMF tests were performed on Alloy 800 which
incorporated either tensile or compressive hold times of 5 minutes.

A concern which should be addressed concerning TMF test technique is
the effect of forced air cooling on thermal stress development within our
test specimen. The question is whether or not preferential cooling of the
specimen surface causes enhanced tensile stresses, resulting in premature
crack initiation. The thermal diffusivity and thermal conductivity of both
alloys are approximately 4.1 mm²/s and 5.1 cal/s·m·°C, respectively, over
the temperature range of interest. We assume convective cooling of a semi-
infinite rod (15) to approximate our 6.25 mm diameter specimens from 649°C
to 375°C in 30 s (our infrared pyrometry readings should be true surface
measurements). The Fourier number for this test specimen is 12.6, which at a
relative temperature of 0.52 (400°C surface temperature) corresponds to a
Biot number of 0.02-0.05. The resulting temperature difference from center
to surface is negligible - certainly not large enough to have a significant
effect on surface stresses.

Microstructural Examination

Transmission electron microscopy was used to characterize the
substructures of both alloys. Transverse slices about 0.5 mm thick were cut
from the gage sections of the fatigue specimens using a low speed saw.
Disks 3 mm in diameter were cut from these slices by electrical discharge
machining. Individual disks ground to about 0.1 mm thickness were polished

to perforation with a twin jet electropolisher. The electrolyte used was a solution of 10% perchloric acid and 90% methanol held at <-40°C. The foils were examined in a JEM 200CX operated at 100 kV. Finer precipitates (<100 nm in diameter) were generally distinguished by dark field microscopy and selected area diffraction was used to crystallographically identify the phases.

Fractography

Fracture surfaces examined in the scanning electron microscope were first coated with gold-palladium by vacuum sputtering to eliminate charging effects caused by oxides which formed at elevated temperatures. For fatigue striation spacing measurements, surfaces were carefully oriented normal to the electron beam, and micrographs were taken at crack length intervals measured by the stage micrometer. Average striation spacings at a crack length were determined by counting the number of striations along a line length on the micrographs. Several measurements were taken on each micrograph, and these were averaged to gain the overall average.

Results

Mechanical Testing

Cyclic Hardening. The cyclic hardening of 316 Stainless Steel and Alloy 800 are summarized in Figures 3 through 7. The isothermal cyclic hardening response of both materials is nearly identical, showing roughly balanced cyclic stress peaks throughout the tests and saturation in stress range early in the tests. Decreasing strain rate for 316 Stainless Steel results in an expanded stress range during initial cycling (Figure 3). By 100 cycles, this difference has essentially disappeared and the two tests nearly overlay for most of the test duration.

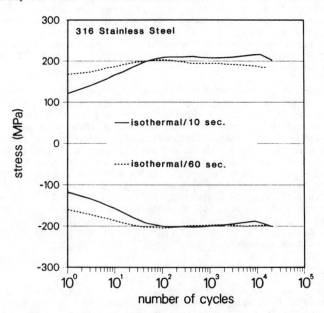

Figure 3. Cyclic hardening of 316 Stainless Steel in isothermal fatigue, 649°C, Δε=0.5%.

For Alloy 800, virtually no difference is observed in the initial isothermal cyclic hardening response for the two strain rates examined. At approximately 1000 cycles, however, the slower test (Figure 4) shows stress saturation and subsequent strain softening, while the faster test continues to slightly harden to failure.

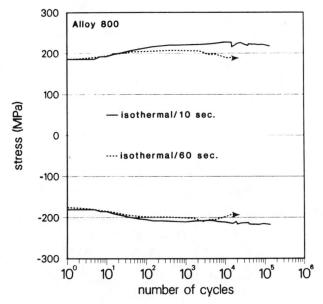

Figure 4.   Cyclic hardening of Alloy 800 in isothermal fatigue, 649°C, Δε=0.5%.

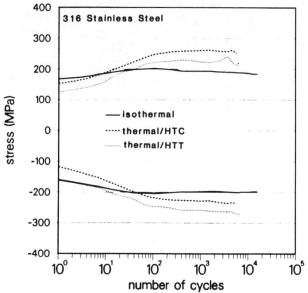

Figure 5.   Cyclic hardening of 316 Stainless Steel in thermomechanical fatigue.

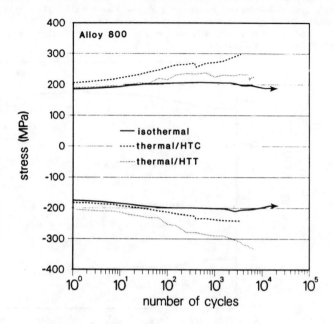

Figure 6. Cyclic hardening of Alloy 800 in thermomechanical fatigue.

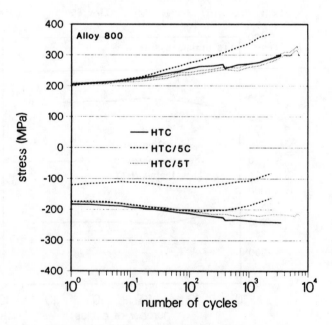

Figure 7. Cyclic hardening of Alloy 800 in thermomechanical fatigue with hold periods.

A major difference in the cyclic hardening response of these two alloys appears during TMF testing. Under thermomechanical cycling, 316 Stainless Steel displays a hardening response which is qualitatively similar to that it shows during isothermal testing (Figure 5). In both cases, cyclic stress peak saturation is reached in about 100 cycles. However, the total stress range during thermomechanical cycling is about 25% larger than in isothermal tests. Stress peaks in both tension and compression are increased, with the largest difference occurring on the low temperature side of the cycle; the result is the development of 20-30 MPa mean stresses during TMF.

In contrast, the TMF cyclic hardening response of Alloy 800 is qualitatively unlike that developed for isothermal testing. The stress peak histories are distinctly asymmetric in shape, with no stress peak saturation on the low temperature side of the cycle. Again the stress peak histories are asymmetric in shape, resulting in the development of roughly 30 MPa mean stresses (Figure 6). These mean stresses are the result of low temperature stress peaks which are approximately 60 MPa higher than observed in isothermal testing, while the high temperature stress peak is essentially coincident with that observed in isothermal testing. Incorporation of a five minute compressive hold period during HTC cycling (Figure 7) results in significantly enhanced tensile mean stress development (~100 MPa) while a five minute tensile hold yields results which are virtually indistinguishable from the continuous cycling test.

Table 2 - Plastic Strain Ranges in Isothermal and Thermomechanical Tests.

Material	Test Type	Dep (%)
316 Stainless	isothermal, 649°C	0.25
316 Stainless	thermal/HTC	0.19
Alloy 800	isothermal, 649°C	0.25
Alloy 800	thermal/HTC	0.15
Alloy 800	thermal/HTC/5C	0.13
Alloy 800	thermal/HTC/5T	0.13

Measurements of plastic strain range in the TMF tests were made at the zero stress values of the stress-temperature hysteresis loops. At zero stress, the difference in temperatures on opposite sides of the loop cause a thermal strain which is purely plastic ($\alpha\Delta T = \Delta\sigma/E = 0$). Measured values are given in Table 2. Since stress ranges for the TMF tests are higher than for the isothermal tests, the elastic strain ranges are necessarily greater. With equal total strain ranges, the plastic strain ranges for the TMF tests are significantly lower than for the isothermal tests.

Fatigue Life. The number of cycles to failure for the specimens examined in this study are listed in Table 3, where all results are for 0.5% total strain range. The cyclic life of 316 Stainless Steel was found to be moderately sensitive to both strain rate and thermomechanical history over the range of strain rate/temperature examined here. In isothermal cycling, increasing the cycle time by a factor of six (from 10 to 60 s) decreases the fatigue life to 75%. For the same 60 second cycle time, TMF cycling has a more dramatic effect - a reduction in life to 40%. Interestingly, this decrease in life is similar for both the HTC and HTT thermomechanical cycles.

161

Table 3 - Summary of Cycles to Failure, 0.5% Total Strain Range

Material	Test Type	Strain Rate (s-1)	Cycles to Failure
316	isothermal, 649°C	0.0010	22000 (2)
316	isothermal, 649°C	0.00017	16700 (1)
316	thermal/HTC	0.00017	5600 (3)
316	thermal/HTT	0.00017	6900 (2)
800	isothermal, 649°C	0.0010	116000 (4)
800	isothermal, 649°C	0.00017	>15900
800	thermal/HTC	0.00017	5700 (2)
800	thermal/HTC/5C	0.00017	2400 (2)
800	thermal/HTC/5T	0.00017	7300 (1)
800	thermal/HTT	0.00017	6900 (2)

( ) - Number of specimens tested for each condition

Alloy 800 displays an much stronger effect of TMF. The cyclic lives of specimens thermomechanically fatigued at 60 s/cycle are only about 5% of the average value of samples isothermally fatigued at 10 s/cycle at 649°C. The fraation of this life reduction due to rate sensitivity was not completely established, but an isothermal test using the slower 60 s/cycle time was run to 15900 cycles without any evidence of impending failure. This is significantly longer than any of the TMF tests. Samples cycled with HTC thermomechanical cycles and 5 minute compressive hold periods (5C) show the shortest cyclic lives.

Comparing the two alloys, the LCF cyclic life of Alloy 800 is about 5 times that of 316 stainless; the lives measured here agree well with published compilations (5, 6). TMF cycles to failure are nearly equal for HTC cycling of the two alloys.

Microstructural Observations

316 Stainless Steel. Isothermal cycling at 0.5% strain range and 649°C produces the changes from the solution annealed structure shown in Figure 8 which contrasts with the nearly dislocation free microstructure of the starting condition. This strain range produces a highly heterogeneous dislocation substructure with most regions near grain boundaries having developed a cell structure with cell sizes of about 1 μm. In some areas, these cells are narrow and well ordered while in other areas the walls are more loosely knit. The centers of almost all grains, and the whole of some grains, did not develop a cellular dislocation structure.

Both intergranular and transgranular precipitation of $M_{23}C_6$ occurred during this test which lasted only 36 hours. Carbides formed intergranularly in the form of blocky, 0.1 to 0.2 μm particles which have formed in sufficient density to be nearly continuous along the boundaries. Also present was a dispersion of very fine (<50 nm) carbide particles distributed throughout the substructure.

The substructure developed during thermomechanical cycling is shown in Figure 9. The dislocation structure produced in this test was similar in character to that formed isothermally with differences in only the volume fraction of cellular structure. Most cells which did form were not well developed and were about 1 μm in size.

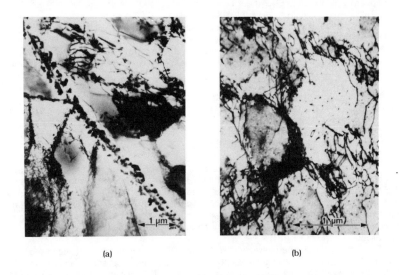

(a)   (b)

Figure 8. Transmission electron micrographs of
316 Stainless Steel isothermally fatigue cycled at
649°C, 6 s/cycle, 36 hr. test duration.

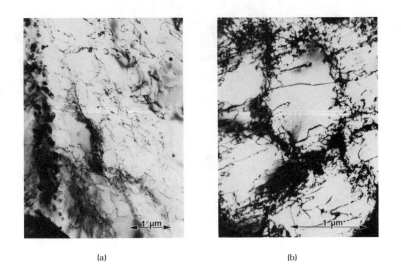

(a)   (b)

Figure 9. Transmission electron micrographs of
316 Stainless Steel thermomechanical HTC cycled,
60 s/cycle, 117 hr. test duration.

The thermomechanically cycled test lasted 117 hours and also showed $M_{23}C_6$ precipitation. Particles about 0.1 μm in size had formed on grain and twin boundaries with the density of coverage lower than observed in the isothermal test. We found no evidence of intragranular carbide precipitation in this TMF test.

Alloy 800. The microstructures shown in Figure 10 were taken from an isothermal test conducted at a slow strain rate of 1.7×10^{-4} s-1 and run for 270 hours. Much of the material showed a tendency toward cell formation. Where they were observed, these cells tended to be several micrometers in size.

The $M_{23}C_6$ carbide was present in the starting material and continued precipitation had occurred during testing. In addition to the idiomorphic 0.5 - 1.0 μm particles already present, our examination showed intergranular precipitation of blocky particles 0.1-0.2 μm in size. Also present was a very fine dispersion of $M_{23}C_6$ particles shown in Figure 10a.

The sample of continuous cycling TMF examined was in test for only 75 hours which is reflected in the substructure. Figure 11 shows that no additional $M_{23}C_6$ precipitation was observed. We found clear evidence of cell formation although it was not extensive and these cells ranged in size from 1- 2 μm. The cell formation was more extensive in the thermomechanically cycled specimen than in the isothermally cycled specimens.

The specimens thermomechanically cycled with hold periods were also examined by TEM and characteristic substructures are shown in Figure 12. Most regions of the materials showed some cell formation with many cells having well-ordered dislocation arrays in the walls. These cells tended to be about 0.5 μm in size and, in general, smaller than those developed under continuous thermomechanical cycling conditions.

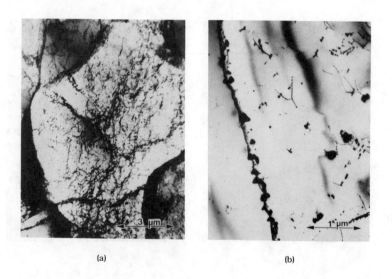

(a)   (b)

Figure 10. Transmission electron micrographs of
Alloy 800 isothermally cycled at 649°C, 60 s/cycle,
256 hr. test duration.

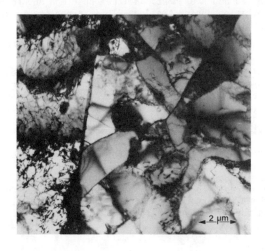

Figure 11. Transmission electron micrograph of
Alloy 800 thermomechanical HTC cycled, 60 s/cycle,
75 hr. test duration.

(a)   (b)

Figure 12. Transmission electron micrographs of
Alloy 800 thermomechanical HTC cycled, 60 s/cycle:
(a) 5 min. tensile hold, 120 hr. test duration;
(b) 5 min. compressive hold, 40 hr. test duration.

165

The test with a 5 minute compressive (and maximum temperature) hold period failed at 2380 cycles after 40 hours of test and the test with a 5 minute hold period in tension (at the minimum temperature) failed at about 7200 cycles or 120 hours of test. Neither specimen showed any indication of carbide precipitation during testing. This is reasonable given the short time either specimen was exposed to elevated temperature.

Fractographic Observations

Isothermal Testing. Fracture surfaces of 316 Stainless Steel isothermally cycled at 649°C are shown at two different crack lengths in Figure 13. Crack growth occurs by ductile striation formation over the entire surface, with considerable secondary cracking out of the main crack plane also evident. At short crack lengths (<0.75 mm) the surface is obscured by oxidation, and striations are not visible.

Fatigue fracture of Alloy 800 at the same temperature shown in Figure 14 is significantly different. Here cracking proceeds by a mixed mode which is partially intergranular and partially ductile. Blocky features on the surface are comparable to the grain size of 10 - 15 μm. Some striations are visible only over a limited crack length range of 1.0 - 1.6 mm, and these are mixed with the intergranular features.

(a)

(b)

Figure 13. Fatigue fracture surface of 316 Stainless Steel isothermally cycled at 649°C: (a) crack length = 1.33 min; (b) crack length = 1.97 mm.

166

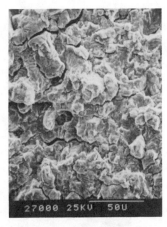

Figure 14. Fatigue fracture surface of Alloy 800
isothermally cycled at 649°C: (a) crack length =
1.30 mm; (b) crack length = 2.02 mm.

Thermal Fatigue. Cracking of both alloys cycled under HTC conditions
occurs by the ductile striation mechanism. Figure 15 shows continuously
cycled Alloy 800 and 316 Stainless Steel. Roughly spherical particles on
the surface of the Alloy 800 sample are nucleated surface oxide; these grow
larger and become closer together at short crack lengths. HTC
thermomechanical fatigue tests of Alloy 800 with either 5 minute compressive
(5C) or tensile (5T) hold periods also crack by striation formation (Figures
16 and 17).

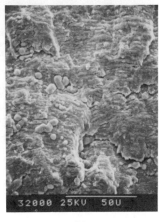

Figure 15. Fatigue fracture surface of HTC
thermomechanical fatigue cycle specimens:
(a) Alloy 800, crack length = 1.00 mm;
(b) 316 Stainless Steel, crack length = 2.22 mm.

167

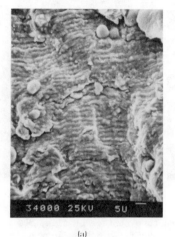 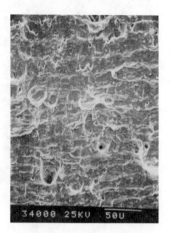

(a) (b)

Figure 16. Fatigue fracture surface of HTC
thermomechanical fatigue cycled Alloy 800
with 5 min. compressive hold period: (a)
crack length = 0.77 mm; (b) crack length =
2.00 mm.

 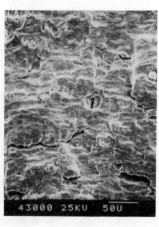

(a) (b)

Figure 17. Fatigue fracture surface of HTC
thermomechanical fatigue cycled Alloy 800
with 5 min. tensile hold period: (a) crack
length = 0.66 min; (b) crack length = 2.84 mm.

HTT thermomechanical fatigue produced mixed mode failures in both alloys. For 316 Stainless Steel (Figure 18a), the surface is comprised of several large regions (~0.3 mm diameter) of intergranular failure which are interconnected by areas of ductile tearing. No striations are visible. Alloy 800 shows similar behavior, except the size scale is different. A few predominantly intergranular islands visible are about 0.1 mm in diameter (Figure 18b) and are separated by ductile tearing with void formation. Again, no striations are visible.

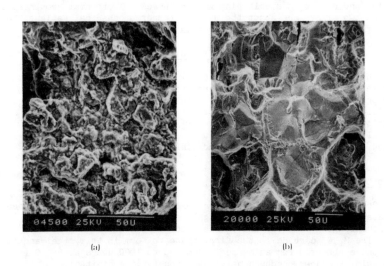

(a)   (b)

Figure 18. Fatigue fracture surfaces of HTT themomechanical cycled specimens showing regions of mixed intergranular and intragranular ductile fracture: (a) Alloy 800; (b) 326 Stainless Steel.

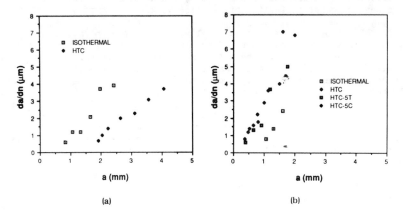

(a)   (b)

Figure 19. Striation spacing crack growth rates versus crack length for isothermal and HTC thermomechanical cycling for (a) 316 Stainless Steel and (b) Alloy 800.

<u>Crack Growth Rates</u>. Fatigue crack growth rates measured by striation spacings are shown in Figure 19a and b for the test conditions where distinct striations could be observed. Isothermal tests with long cycle times were too oxidized to view striations, and the HTT thermomechanical fatigue tests did not form striations. With 316 Stainless Steel, growth rates are significantly higher for isothermal cycling than for thermomechanical fatigue HTC loading. Contrary to this, TMF cycling in Alloy 800 produces a higher growth rate than isothermal cycling. The three sets of data for HTC cycling with and without hold times in Alloy 800 are very similar and result in overlapping growth curves. Isothermal growth rates for the two alloys are about the same, but thermomechanical fatigue HTC crack growth in Alloy 800 is faster than in 316 Stainless Steel.

Discussion

The large decrease in cycles-to-failure for TMF is an observation of sizable engineering impact which warrants rationalization and interpretation. The discussion which follows is aimed at examining three possible sources for these observations:
- unique microstructural changes,
- cyclic stress-strain response,
- and their effects on fatigue crack initiation and propagation.

Microstructural Observations

Precipitation of $M_{23}C_6$ carbides during elevated temperature exposure of 316 Stainless Steel has been thoroughly examined under conditions of isothermal aging (11, 12) and in elevated temperature low cycle fatigue (16, 17). The $M_{23}C_6$ carbide is the first phase to form in this steel and is the only phase observed for aging periods of up to 1000 hours at 650°C (11). This aging produces predominantly grain boundary precipitation of large particles and no intragranular precipitation when the starting material was well annealed. Heterogeneous nucleation becomes predominant during low cycle fatigue at elevated temperature (16, 17). These particles form in a distribution reflective of the current dislocation substructure and would essentially lock-in that structure, but the density of these precipitates is not sufficient to alter the flow stress of the alloy (17). The results of this study are consistent with this earlier work.

The effect of thermomechanical cycling on this precipitation process (Figure 9) is to simply retard the formation rate of $M_{23}C_6$, and no unique synergism appears to occur. The TTT (time-temperature-transformation) curve for this reaction (11) indicates that the nose of the curve is at about 800°C so that the thermomechanical cycling in this study encompassed a range fully below the nose of the curve. Lai (12) has shown that the rate of $M_{23}C_6$ precipitation is chromium diffusion controlled in this region. This would suggest that the effect on precipitation kinetics of thermomechanical cycling between 360 and 650°C would only appear through the temperature dependence of chromium diffusivity. Our results are consistent with this rationale. Thermomechanical cycling which extends to higher temperatures above the nose of the TTT-curve may, however, produce different effects. This is a possibility since the controlling mechanism in $M_{23}C_6$ precipitation would be changing over the period of each thermomechanical cycle: from nucleation controlled at the peak temperature to growth controlled by Cr diffusion at the lower temperatures.

The mechanical strain range was 0.5% in both the isothermal and the thermomechanical cycling tests so that differences in plastic strain range and the dislocation substructures would reflect only the thermal cycling.

170

As Figures 8 and 9 indicate, the differences are minor. The cell walls in the isothermally aged specimens did appear to be generally better ordered, but not uniformly so. There appear to have been competing processes which have combined to produce these fairly similar substructures. The dislocation substructure development for LCF is hindered by the precipitation of the intragranular $M_{23}C_6$ particles. Because of this distribution of hard precipitate particles, a well developed subgrain structure cannot develop during continuous cycling even though the temperature is high enough to facilitate the necessary climb to produce a well ordered subgrain structure. The limiting features under thermomechanical cycling conditions are different. Since dislocation climb is more difficult at the lower temperatures, thermomechanical cycling would not be expected to give rise to a well ordered substructure. Climb and dislocation annihilation processes cannot proceed at the cool end of the thermomechanical cycles so that dislocation tangling and multiplication would occur. The final structure then reflects some climb and annihilation by dislocations at the higher temperatures but also reflects the dislocation multiplication and tangling going on at lower temperatures.

For Alloy 800, as with 316 Stainless Steel, precipitation of $M_{23}C_6$ is the dominant precipitation process at temperatures below 650°C. Comparison of Figures 10 and 11 show microstructural evolution effects similar to those just discussed for 316 Stainless Steel. Intragranular precipitation similar to that shown in Figure 10 of $M_{23}C_6$ in Alloy 800 during elevated creep and fatigue has been observed earlier (8, 13, 18). The approximate TTT-curve given by Jones (9) for $M_{23}C_6$ formation indicates a maximum precipitation rate at about 720°C which is above the peak temperature used in this study. The available carbon in the starting matrix of this heat of material is sufficient to produce subsequent precipitation, both intragranularly and intergranularly, in the isothermally cycled specimen (Figure 10). This precipitation has not occurred in the thermomechanically cycled specimen (Figure 11) during the time of the test. Here again, the precipitation rate is controlled by diffusion, probably chromium, and the thermomechanical cycling slows the precipitation kinetics accordingly. Had the thermomechanical cycling test continued, the $M_{23}C_6$ carbide would have formed after longer times. Similar arguments apply to the microstructures developed in thermomechanical cycling with hold periods shown in Figure 12. The total time at high temperature was insufficient to result in $M_{23}C_6$ formation.

The carbide precipitation in the isothermal test of Alloy 800 appears to have influenced the dislocation substructure development. This can be seen by comparing Figures 10 and 11. As was observed in 316 Stainless Steel, the fine distribution of carbides hindered the development of a clean cell structure. The thermomechanically cycled material showed a cleaner cell structure in spite of the large fraction of time spent at lower temperatures. The specimen which was subjected to a 5 minute hold period at the peak temperature (compressive strain peak) developed the clearest subgrain structure with narrow, low angle boundaries. The specimen in Figure 12 was held at the minimum temperature for 5 minutes and the dislocation substructure is very similar to that formed during continuous cycling TMF. These are consistent with the time-temperature histories of those tests.

The γ' phase was not observed in any of the Alloy 800 specimens, although this heat of Alloy 800 is known to form γ' under isothermal aging at 649°C (9). An earlier study (8) involving isothermal cycling at 565°C, where a larger volume fraction of γ' is stable, did show its formation during cycling. Fewer and extensively cut γ' particles were present for the LCF history than for the isothermal aging. The results at 649°C from this study indicate that the precipitate cutting followed by resolution which occurs during cycling effectively precludes Y' formation during any kind of mechanical cycling.

In summary, no unique synergism between TMF and microstructural development appears to have occurred for either 316 Stainless Steel or Alloy 800. Both the carbide precipitation and dislocation substructure evolution reflect the slowing of both solute and vacancy diffusion which takes place as the temperature is decreased below 649°C. Nothing was observed to indicate that either carbide morphology or dislocation substructure differences contribute significantly to the difference in cycles-to-failure observed between LCF and TMF histories.

Cyclic Hardening

Two principal observations may be made regarding the cyclic hardening response of 316 Stainless Steel and Alloy 800: both alloys develop mean stresses during thermomechanical cycling and both quantitative and qualitative differences exist in the cyclic development of these mean stresses. Two separate macroscopic mechanisms may be invoked to explain the continuous cycling stress asymmetry, the combined effect of which may also explain the differing response of the two alloys.

The first contribution to mean stress development is the influence of temperature on the plastic constitutive relationships for these alloys. Both 316 Stainless Steel and Alloy 800 exhibit a dependence of yield stress and cyclic strain hardening within the temperature range of our TMF tests (19), the result of which is, in general, a slight decrease in flow stress with increasing temperature. As part of a companion study of 304 stainless steel (20), calculations simulating TMF have been performed using a temperature dependent, elastic-plastic constitutive model. It has been shown that TMF of a material displaying such temperature dependent deformation can result in the development of mean stresses on the low temperature side of the cycle, as is shown by both 316 Stainless Steel and Alloy 800. Thus mean stress development can be explained by evoking a simple dependence of plastic flow on temperature.

The second contribution to mean stress development is the character of viscous flow in these two alloys. Previous work can be examined to conclude that time dependent flow is more important during cyclic loading for Alloy 800 than for 316 Stainless Steel. Isothermal low cycle fatigue tests have been performed by Jones (21) at 649°C with 5 minute compressive holds for both alloys. It was seen that Alloy 800 shows approximately three times the level of stress relaxation that was displayed by 316 Stainless Steel. Also, Alloy 800 develops significant tensile mean stresses during hold time cycling, while 316 Stainless Steel does not. Thus one can expect that Alloy 800 would show more structural relaxation during the high temperature end of the TMF cycle than 316 Stainless Steel, resulting in enhanced mean stress development on cooling and reversed straining. This is supported by the observation that incorporating a five minute compressive hold in the HTC thermomechanical fatigue testing of Alloy 800 results in exaggerated mean stress development similar to that shown in continuous cycling.

The combination of these two effects would lead to mean stress development in both alloys with an exaggerated effect in Alloy 800, which is consistent with the observations of this study. We have not determined the micromechanism by which these two alloys differ. It can be deduced that the differences are subtle since the substructures are so similar. This issue is the subject of continuing work. In summary, the development of significant tensile mean stresses under TMF conditions would be expected to

172

contribute to a loss of fatigue life through either or both crack nucleation or growth.

Fractography

The fatigue fracture modes of this category of austenitic alloys are well known to be temperature and strain rate dependent (22, 23, 24). At lower temperatures or high rates, propagation by ductile striations dominates, while at elevated temperatures intergranular cracking will occur at slow strain rates or with tensile hold periods. Results for the isothermal tests at 649°C and 10 s/cycle show that 316 Stainless Steel is still in the ductile regime while Alloy 800 is mixed mode with combined ductile striation and intergranular cracking. For HTC thermomechanical cycling where the peak tensile loads are at the low temperature extreme of the thermomechanical cycle (360°C), crack advance occurs at low temperature by ductile striation formation. Thus for Alloy 800 the mode is shifted to the ductile mode even for the hold period tests.

In the HTT tests of both alloys, maximum tensile loading and crack advance occur at the peak temperature of the thermomechanical cycle. Both alloys failed in mixed mode where intergranular fracture formed as island-like regions separated by ductile tearing. This suggests that the lower strain rate of these 60 s/cycle tests promotes intergranular fracture and that the intergranular regions nucleated ahead of the main crack front. This mode change did not significantly affect the total cycles to failure for either alloy.

Fatigue Life

It is possible to calculate the number of cycles spent in propagating a fatigue crack across each specimen by integrating the crack growth data given in Figure 20. This was done using a numerical approximation where the growth rate between any two adjacent crack length intervals was taken as the average of the rates at the beginning and end of the interval. Final crack lengths were measured for each sample, and an initial growth rate equal to the measured rate at the shortest observable crack length was assumed from initiation to that length. This last assumption may cause the calculation to somewhat underestimate the number of propagation cycles for small crack lengths. Errors of even 150% in this calculation would not affect the overall conclusions.

Calculated values of cycles spent in crack propagation and, by subtraction, cycles for crack initiation are given in Table 4 along with values of total fatigue life for the particular specimens examined. For 316 Stainless Steel in isothermal cycling, about 15% of the fatigue life is spent in crack propagation, requiring 85% for the initiation process. This fraction is similar to previous measurements at an equivalent strain range at 593°C on 316 Stainless Steel (25) and 304 Stainless Steel (26). Isothermal cycling in Alloy 800 did not yield striations over enough of the crack length to allow the propagation calculations, but the limited rates measured (Figure 20b) are comparable to 316 Stainless Steel. Again, earlier measurements of Alloy 800 at 565°C and 0.5% total strain range showed propagation by ductile striations and approximately 90% of life spent in crack initiation (27). This would imply that the propagation cycles here are a small fraction of the total for isothermally cycled Alloy 800.

Table 4 - Cycles for Crack Propagation and Total Cycles to Failure

Material	Test Type	Cycles in Propagation	Total Cycles	Cycles in Initiation*
316	isothermal, 649°C	3100	21514	18400
316	thermal/HTC	4300	7060	2800
800	isothermal, 649°C	---	110000	~100000
800	thermal/HTC	1100	4500	3400
800	thermal/HTC/5C	1000	2418	1400
800	thermal/HTC/5T	1500	7300	5800

*Approximate based on $N_t = N_i + N_p$

Crack initiation arguments can be used to explain the large difference in fatigue life between the two alloys in isothermal cycling (Table 3). Taira, et al (28) showed that cycles to microcrack initiation decreases with increasing grain size for a ferritic low carbon steel. Essmann et al (29) and Mughrabi et al (30) relate this to diminished surface intrusion and extrusion formation for smaller grain material. Intrusions and extrusions form as a result of localized dislocation slip intersecting the sample surface as persistent slip bands (PSB's); microcracks can initiate at PSB's with large height or slip offset. The height of PSB's decreases with shorter slip distance to the nearest grain boundary, so PSB height and microcrack formation are less for small grain material. Cheng and Laird (31) experimentally showed that PSB height increases with plastic strain amplitude in copper crystals and from this derived a model which showed that cycles to initiation decreases with increasing PSB offset. Given that the measured isothermal cyclic stress-strain hardening responses of small grain Alloy 800 and large grain 316 Stainless Steel are nearly identical, leading to equal plastic strain ranges, the grain size effect on crack initiation likely accounts for the large difference in total isothermal life of the alloys.

Fractions of life spent in propagation during HTC cycling range from 20 to 40% for Alloy 800 samples to 60% for 316 Stainless Steel (Table 4). These relative fractions are much greater than for isothermal cycling at 649°C. Cycles for propagation of Alloy 800 for continuous and 5 minute hold period thermomechanical fatigue tests are similar, within the uncertainties of the method. The large drop in life for 316 Stainless Steel is accompanied by an equal or greater number of cycles for propagation.

These data clearly show that crack propagation effects also cannot be used to explain the large decreases in fatigue life observed with TMF cycling, since propagation does not dominate the total life at this strain range. By deduction, this study suggests that crack initiation also dominates the TMF response. One possibility was that the experimental technique itself might give rise to surface thermal stresses leading to early crack initiation. The analysis described in the Experimental Procedures section shows this to be unlikely.

The initiation process in TMF cycling occurs under significantly different mechanical and thermal conditions from the isothermal case. In addition, results also indicate that the grain size differences between the two alloys does not significantly influence failure rates. Much of the plastic deformation in the loading cycle takes place at lower temperatures, where it is expected that athermal deformation processes would dominate. This could tend to limit thermally activated climb and cross-slip processes, thus promoting more planar slip which would favor PSB formation. TEM surveys of the bulk did not show a strong tendency toward planar slip, and closer examination of surface deformation features is required. As shown earlier, stress range is greater and plastic strain range is smaller in TMF cycling. Cheng and Laird's model for initiation (31) based on PSB formation predicts that lower strain range would lead to delayed crack initiation, which was not observed here. This suggests that higher stress, independent of plastic strain range, is significant, but at this time it is difficult to completely assess its effects. Higher stresses could increase the number of dislocations piled up at grain boundaries, allowing more offset at the surface. Conversely, higher resolved stress on slip planes could lead to increased cross-slip and more diffuse slip lines; this would favor less concentrated PSB's and delayed initiation. To sort out these possible effects, additional work is continuing to directly observe the initiation processes and relate them to near surface deformation features during TMF cycling.

Conclusions

1. TMF between 649 and 360°C shortens the fatigue life of both 316 Stainless Steel and Alloy 800 when compared to LCF at 649°C at a strain range of 0.5%. This loss of fatigue life is more marked in Alloy 800 than for 316 Stainless Steel. In both alloys, this effect must be considered in the design of structures which see thermal cycling in service.

2. This TMF cycling does not produce any unique synergisms in either the precipitate or dislocation substructures. The $M_{23}C_6$ carbide formed in both alloys during LCF but not during TMF for the duration of testing in this study. Dislocation cell structures were incompletely developed in the LCF samples due to intragranular $M_{23}C_6$ precipitation and in the TMF samples due to the difficulty of recovery at the low temperature end of each cycle.

3. Mean stresses are developed by both alloys under TMF conditions. These are larger for Alloy 800 than for 316 Stainless Steel with the HTC cycle producing tensile mean stresses. Differences in time-dependent flow characteristics between these alloys produce this effect.

4. Fractographic analysis has shown that TMF accelerates fatigue crack growth in Alloy 800 and may slightly retard crack growth in 316 Stainless Steel. In both alloys, cracking occurs by a ductile striation mechanism. Under LCF conditions, 316 Stainless Steel showed ductile striations while Alloy 800 showed a mixture of intergranular and ductile intragranular modes.

Acknowledgements

The authors acknowledge the assistance of D. T. Schmale in the development of the TMF test system and of C. R. Hills in the transmission electron microscopy.

References

1. H. Sehitoglu, "Characterization of Thermomechanical Fatigue," pp 93- 110 in Thermal and Environmental Effects in Fatigue: Research/Design Interface, ASME, New York, 1983.

2. J. Ginsztler, "Assessment of Thermal Fatigue Resistance of Some Boiler Steels," pp 335-338 in Pressure Vessel Technology, Vol. 1, Inst. of Mech. Engrs., London, 1980.

3. D. J. Marsh, "A Thermal Shock Fatigue Study of Type 304 and 316 Stainless Steels," Fatigue Eng. Mater. Struct., 4 (1981) pp 179-195.

4. J. B. Conway, J. T. Berling, and R. H. Stentz, "Low-Cycle Fatigue and Cyclic Stress-Strain Behavior of Incoloy 800," Met. Trans, 3 (1972) pp 1633-1637.

5. D. R. Diercks, "A Compilation of Elevated-Temperature, Strain-Controlled Fatigue Data on Type 316 Stainless Steel," Argonne National Laboratory Report ANL/MSD-77-8, 1978.

6. S. Majumdar, "Compilation of Fatigue Data for Incoloy 800 Alloys," Argonne National Laboratory Report ANL/MSD-78-3, 1978.

7. J. Orr, "A Review of the Structural Characteristics of Alloy 800," pp 25-60 in Alloy 800, W. Betteridge, et al, eds.; North-Holland Publishing Company, New York, 1978.

8. W. B. Jones and R. M. Allen, "Mechanical Behavior of Alloy 800 at 838 K," Met. Trans. A, 13A (1982) pp 637-648.

9. W. B. Jones, "Creep-Fatigue and Temperature Synergisms in Alloy 800," pp 403-418 in Fracture: Interactions of Microstructure, Mechanisms and Mechanics, J. M. Wells and J. D. Landes, eds., AIME, New York, 1984.

10. A. H. Nahm and J. Moteff, "Characterization of Fatigue Substructure of Incoloy Alloy 800 at Elevated Temperature, Met. Trans. A, 12A (1981) pp 1011-1025.

11. B. Weiss and R. Stickler, "Phase Instabilities During High Temperature Exposure of 316 Austenitic Stainless Steel," Met. Trans., 3 (1972) pp 851-865.

12. J. K. L. Lai, "A Study of Precipitation in AISI Type 316 Stainless Steel." Mater. Sci. and Eng., 58 (1983) pp 195-209.

13. H. Nahm and J. Moteff, "Second Phase Formation and its Influence on the Fatigue Properties of Incoloy 800 at Elevated Temperatures," Met. Trans A., 7A (1976) pp 1473-1477.

14. D. T. Schmale and R. W. Cross, "Techniques for Elevated Temperature Mechanical Testing," Sandia National Laboratories Report SAND82-0727 (NUREG/CR-2793, 1982.

15. M. D. Mikhailov and Ozisik, Unified Analysis and Solutions of Heat and Mass Diffusion, pp 305-318, Wiley Interscience Publishers, New York, 1984.

16. K. D. Challenger and J. Moteff, "Correlation of Substructure with the Elevated Temperature Low-Cycle Fatigue of AISI 304 and 316 Stainless Steels," in Fatigue at Elevated Temperatures, ASTM STP 520, A. E. Carden, et al, eds., ASTM, Philadelphia, 1973.

17. E. R. De Los Rios and M. W. Brown, "Cyclic Strain Hardening of 316 Stainless Steel at Elevated Temperatures," Fatigue Eng. Mater. Struct., 4 (1981) pp 377-388.

18. A. Orlova, K Milicka, and J. Cadek, "Precipitation of Intragranular $M_{23}C_6$ Particles and their Effect on the High Temperature Creep of Austenite," Mat. Sci and Eng., 50 (1981) pp 221-227.

19. P. S. Maiya, "Cyclic Stress-Strain Curves and Cyclic-Hardening Behavior for Types 304 and 316 Stainless Steel and Incoloy Alloy 800H," Argonne National Laboratory Report ANL-80-100, 1980.

20. R. J. Bourcier, "Numerical Simulation of the Thermal Fatigue of 304 Stainless Steel," presented at 1986 AIME Fall Meeting, Orlando, FL, October 1986.

21. W. B. Jones, unpublished research, 1987.

22. D. J. Michel and H. H. Smith, "Accelerated Creep-Fatigue Crack Propagation in Thermally Aged Type 316 Stainless Steel," Acta Metallurgica, 28 (1981), pp 999-1007.

23. S. Taira, R. Ohtani, and T. Komatsu, "Application of J-Integral to High-Temperature Crack Propagation, Part II -- Fatigue Crack Propagation," Journal of Engineering Materials and Technology, Vol. 101, pp 162-167.

24. J. Wareing, H. G. Vaughan, and B. Tomkins, "Mechanisms of Elevated-Temperature Fatigue Failure in Type 316 Stainless Steel," UKAEA ND-R-447(S), September 1980.

25. J. A. Van Den Avyle, "Low Cycle Fatigue of Tubular Specimens," Scripta Metallurgica, Vol. 17, pp 737-740.

26. P. S. Maiya, in "Mechanical Properties Test Data for Structural Materials, Quarterly Progress Report for Period Ending January 31, 1975," ORNL-5104, 1975, p 13.

27. J. A. Van Den Avyle, "Fatigue Crack Growth Under Fully Plastic Cycling in 316 Stainless Steel and Alloy 800," to be published.

28. S. Taira, K. Tanaka, and M. Hoshima, "Grain Size Effect on Crack Nucleation and Growth in Long-Life Fatigue of Low-Carbon Steel," Fatigue Mechanisms, ASTM STP 675, 1979, pp 135-173.

29. U. Essmann, U. Gosele, and H. Mughrabi, "A Model of Extrusion and Intrusions in Fatigued Metals; Part 1: Point Defect Production and Growth of Extrusions," Philosophical Magazine A, Vol. 44, 1981, pp 405-426.

30. H. Mughrabi, R. Wang, K. Differt, and U. Essmann, "Fatigue Crack Initiation by Cyclic Slip Irreversibilities in High Cycle Fatigue," Fatigue Mechanisms: Advances in Quantitative Measurement of Physical Damage, ASTM STP 811, 1983, pp 5-45.

31. A. S. Cheng and C. Laird, "Fatigue Life Behavior of Copper Single Crystals, Parts 1 and 2," Fatigue of Engineering Materials and Structures, Vol. 4, 1982, pp 331-341 and 343-353.

BITHERMAL LOW-CYCLE FATIGUE BEHAVIOR

OF A NiCoCrAlY-COATED SINGLE CRYSTAL SUPERALLOY

J. Gayda, T. P. Gabb, R. V. Miner, and G. R. Halford
NASA Lewis Research Center
21000 Brookpark Road
Cleveland, Ohio 44135

ABSTRACT

Specimens of a single crystal superalloy, PWA 1480, both bare
and   coated with a NiCoCrAlY alloy,   PWA 276,   were tested in
low-cycle   fatigue at   650 and   1050C,   and in 'bithermal'
thermomechanical fatigue tests.   In the two bithermal test
types, tensile strain was imposed at one of the two temperatures
and reversed in compression at the other.   In the high-strain
regime, lives for both bithermal test types approached that for
the 650C isothermal test on an inelastic strain basis, all being
controlled by the low ductility of the superalloy at 650C.   In
the low-strain regime, coating cracking reduced life in the 650C
isothermal test.   The bithermal test imposing tension at 650C,
termed 'out-of-phase', also produced rapid surface cracking, but
in both coated and bare specimens.   Increased crack growth rates
also occurred for the out-of-phase test.   Inceased lives in
vacuum   suggested   that   there is   a   large   environmental
contribution to damage in the out-of-phase test due to the 1050C
exposure followed by tensile straining at the low temperature.

INTRODUCTION

Gas turbine engine components are subject to complex local
stress-strain-temperature-time cycles during operation. In
addition to mechanical loads imposed centripitally and by gas
impingement, thermal strains are imposed by temperature
gradients and thermal expansion differences between joined
materials. Damage from such cyclic loading is termed
thermomechanical fatigue, or TMF. Advanced turbine engine
blades are Ni-base superalloys cast as single crystals and
coated in various ways with a layer which is basically more
Al-rich for protection from the environment. Relative to the
base superalloy, these coatings have greater coefficients of
thermal expansion, higher elastic moduli than the [001]
lengthwise axis of a single crystal turbine blade, lower flow
stress, particularly at high temperature, good ductility at high
temperature, but very low ductility at temperatures below about
700C (1). Cracking of the coating is promoted in areas of a
component where it is cycled into tension at low temperatures
and these cracks may then propagate into the superalloy below
(2,3). The performance of a particular coating is further
reduced if its coefficient of thermal expansion is greater than
that of the superalloy, since it is placed in tension simply
upon cooling from high temperatures (3).

The eventual aim of this research is to model failure of coated
single crystal superalloy laboratory specimens during
generalized TMF cycling by treating the specimen as a composite
system and using behavior established for both the superalloy
and bulk coating alloy. A system with the overlay type of
coating has been chosen rather than the diffused-in type to
simplify the problem. At least as they go into service, the
overlay coatings have a uniform composition and microstructure
through most of their thickness. Further, they may be produced
in bulk form by processes such as low pressure plasma spraying
and have microstructures closely resembling those of the thin
coatings deposited on real hardware.

The system chosen for study, Ni-base single crystal superalloy
PWA 1480 and NiCoCrAlY low-pressure-plasma-sprayed overlay
coating PWA 276, was developed by Pratt and Whitney Aircraft.
The mechanical behavior of PWA 1480 crystals from the same lot
studied herein have been described previously for isothermal
monotonic (4) and creep-fatigue loading (5). Comparison of
several studies of the monotonic loading behavior shows there is
considerable variability among different lots of PWA 1480 (4).
The bulk coating alloy PWA 276 produced as thick plates by the
same low-pressure-plasma-spray process has also been studied in
high temperature monotonic (6,7) and fatigue tests (8). DeLuca
and Cowles (9) have studied another heat of PWA 1480 single
crystals with a diffusion-aluminide coating. Both fatigue crack
initiation and propagation rate tests were conducted using TMF
cycles in which mechanical strain and temperature were varied
simultaneously.

The present work describes the life and failure mechanisms of
coated and bare crystals of PWA 1480 in isothermal fatigue at
650 and 1050C and 'bithermal' TMF cycles involving mechanical
straining only at these two temperature ends of the cycles. The
temperature 650C represents about the upper limit of low

temperature behavior in both materials characterized by relatively constant high strength and low ductility. While 1050C is about the upper limit of allowable temperature in turbine engine blades and is in the regime where the coating is extremely weak and ductile. The bithermal test was proposed earlier by one of the present authors (10). Experimentally, it simplifies analysis and control since the mechanical and thermal strains are interposed rather than superimposed. Also, this test offers the possibility of easier understanding of TMF since deformation, at least in the superalloy, occurs only at the two temperature limits. This allows comparison and possible unification with isothermal behavior at the two temperatures. The qualification above is presented because stresses during heating and cooling due to thermal mismatch likely produce some inelastic deformation in the coating at high temperatures before the application of external load.

MATERIALS AND PROCEDURES

Materials. The single crystal superalloy PWA 1480 and coating PWA 276 studied herein were developed by Pratt and Whitney Aircraft. PWA 1480 has the following nominal composition in weight percent: 10Cr, 5Al, 1.5Ti, 12Ta, 4W, 5Co, balance Ni. The alloy contains about 65 v/o of the γ' phase, but no carbides or borides.

The crystals were cast by a commercial vendor as round bars about 21 mm in diameter and 140 mm long. They were solution treated for 4 hr at 1290C before machining. Bars having [001] within 7° of the center line were selected for fatigue specimens. These had a 4.8 mm diameter by 15 mm long reduced section. After machining most specimens were coated by low pressure plasma spraying. The PWA 276 coating composition in weight percent was 20.3Co, 17.3Cr, 13.6Al, 0.5Y, and balance Ni. The coating thickness was about 0.13 mm. Both coated and uncoated specimens were given the coating cycle heat treatment of 1080C for 4 hr and aged at 870C for 32 hr.

The microstructure of the superalloy and coating are represented in Fig. 1. It may be seen that there were several types of non-uniformities in the superalloy. Interdendritic porosity is normal in cast single crystal superalloys, and those studied herein contained about 0.3 v/o. The pore diameter averaged 7 μm with a standard deviation of 5 μm. The interdendritic areas also contained undissolved γ' eutectic nodules a few μm to tens of μm in diameter. These occupied 1-2 v/o of the alloy. Elsewhere, the γ' was in cuboids about 0.6 μm on edge. The PWA 276 alloy is about 50 v/o Ni-based solid solution and 50 v/o NiAl-based intermetallic compound. The low-pressure-plasma-sprayed coating had a grain size of about 1.5 μm and contained 1-2 v/o of pores averaging about 20 μm in diameter. The average surface roughness was 8 μm.

Test Procedures. All fatigue testing was done on 90 KN, servohydraulic, closed-loop test machines, one of which was equipped with a diffusion pumped vacuum chamber. Both machines were equipped with a 5 KW radio frequency induction generator for specimen heating. Closed-loop temperature control was

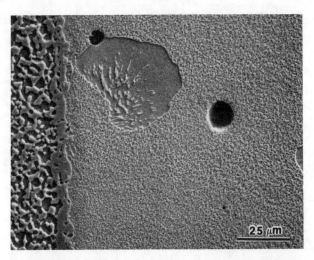

Figure 1. Microstructures of single crystal
superalloy, PWA 1480, and NiCoCrAlY overlay
coating, PWA 276 (left).

employed in all tests, however, two methods of temperature
measurement were used. For tests of coated specimens in air an
infrared pyrometer was used. However, for tests of bare
specimens in air and all tests in vacuum, emissivity was found
to vary significantly with time and a Type K thermocouple was
employed. The junction was pressed against, rather than welded
to, the specimen by wrapping the wires around the opposite side
and applying a slight tension. The thermocouple was calibrated
and periodically re-checked with a disappearing filament optical
pyrometer, which is relatively insensitive to changes in
emissivity. Strain was measured using an axial extensometer
with a 12.5 mm gage length. A dual pen strip chart was used to
record strain/time and load/time data continually, while an x-y
recorder was used to record load/strain hysteresis loops.

Isothermal fatigue tests were conducted at 650 and 1050C under
total mechanical strain control at a frequency of 0.1 Hz. A
digital function generator was employed to produce the
sinusoidal control waveform with an R ratio of -1 (minimum/
maximum strain).

TMF behavior was studied in a simplified bithermal cycle between
650 and 1050C. Specimens were strained at one temperature,
unloaded, changed to the other temperature, and then strained in
the opposite direction to produce a completely reversed strain
cycle. Fig. 2 shows the stress-strain hysteresis loop for what
is termed an 'out-of-phase' cycle in which the tensile and
compressive strains are imposed at the lower and higher
temperatures, respectively. Tensile strain and high temperature
coincide in the 'in-phase' cycle. A 16 bit computer, equipped
with dual digital/analog converters was used to generate the
control waveforms for load and temperature. Total cycle time
was about 120 seconds, of which 100 to 110 seconds elapsed
while changing and stabilizing temperature. The mechanical

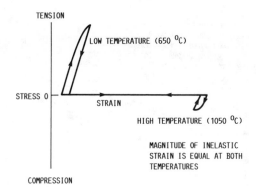

Figure 2. Stress-strain hysteresis loop for an
out-of-phase bithermal test.

strain rate in the bithermal tests was within a factor of
two of that used in the isothermal tests.

For large-strain, short-life bithermal tests, equal tensile and
compressive inelastic strain ranges, $\Delta\varepsilon_{in}$, at the two test
temperatures may be produced by fixing the endpoint sums of the
total mechanical, $\Delta\varepsilon_t$, and thermal, $\Delta\varepsilon_{th}$, strains, Fig. 2.
Under this condition the material will rapidly equilibrate the
tensile and compressive $\Delta\varepsilon_{in}$. A constant $\Delta\varepsilon_t$ test results if
the temperature endpoints, and thus $\Delta\varepsilon_{th}$, are held constant.
However, for small-strain, long-life tests, choosing the
endpoints poses problems. When $\Delta\varepsilon_{in} \leq 10^{-4}$ it cannot be
resolved, and when still lower, the test may be considered
essentially elastic. Unlike for the self-equilibrating
large-strain tests, there are infinite combinations of endpoints
which would produce essentially elastic tests, all with
different maximum tensile and compressive stresses. These tests
would have the equivalent of what would be termed different mean
stresses in an isothermal test. The self-equilibrating
large-strain tests do not. Tests with various endpoints
representing reasonable extrapolations from higher strain range
tests indicated no significant differences in life.

Tests in the low-strain regime were also conducted in load
control with the tensile/compressive load ratio held the same as
in strain-controlled tests at the lower ranges. The rationale
was twofold. First, since loading is nearly elastic, the
constant tensile/compressive load ratio should yield a cycle
nearly 'balanced' in terms of the minute $\Delta\varepsilon_{in}$ as desired in a
strain-controlled test. But second, the stability and accuracy
of the test improves. When the sum of the mechanical and
thermal strains are fixed, fluctuations in the endpoint
temperatures result in small mechanical strain fluctuations, and
in low-strain tests these fluctuations become of the magnitude
of the small $\Delta\varepsilon_{in}$. Control on the maximum tensile and
compressive loads is not similarly affected. Out-of-phase tests
in vacuum were conducted with tensile/compressive load limits
identical to tests run in air. The vacuum was about 10^{-6} Torr.

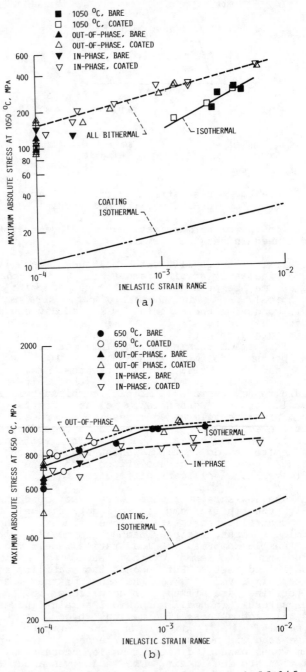

Figure 3. Maximum absolute stress at half life as a function of $\Delta\varepsilon_{in}$ at: (a) 1050C, and (b) 650C

184

Cyclic Stress-Strain Behavior. Fig. 3 compares the cyclic
stress response of the PWA 1480 single crystals in isothermal
and bithermal tests and that of the bulk PWA 276 coating alloy
determined previously in isothermal tests (8). The comparison
is made on the basis of the maximum absolute value of stress at
1050C for a given $\Delta\varepsilon_{in}$ in Fig. 3(a) and similarly for 650C in
Fig. 3(b). The stresses measured at half life are used. The
stress response of the single crystals as a function of the
number of cycles was very stable for the 650C isothermal and
bithermal tests, but about 5-10 o/o softening occured in the
1050C isothermal tests.

It may be seen that the strength of the coating is about 1/10th
that of the superalloy at 1050C, and at 650C it is about 1/3.
Further, the area of a .13 mm thick coating on a 4.8 mm diameter
substrate is about 1/10 of that of the substrate. The stresses
shown for the 1050C tests of the coated single crystals have
been calculated assuming that the coating carried no significant
load. This assumption appears justified by the agreement

(a)   (b)

(c)   (d)

Figure 4. Microstructures of specimens failed
in various tests: (a) 1050C, (b) 650C, (c) In-phase
bithermal, and (d) Out-of-phase bithermal.

185

between the data for the coated and bare single crystals. This was not true at 650C. For the coated specimens, small corrections have been applied to the stresses and inelastic strains in the superalloy for the 650C isothermal tests. As described in the Appendix, it was possible to estimate the load borne by the coating with knowledge of the 650C cyclic stress-strain behavior of the coating and its elastic modulus (8). The load borne by the coating increases the apparent $\Delta\varepsilon_{in}$ at 650C, however in the bithermal tests, the smaller true $\Delta\varepsilon_{in}$ is observed in the 1050C half of the cycle, where the coating bears an insignificant load. Corrections for the 650C isothermal data improved agreement between the cyclic stress-strain behaviors of the bare and coated specimens. Still, corrections were relatively small and did not change any conclusions about the relative life behavior among the various test types.

sRepresentitive ˙transmission electron micrographs of specimens failed in each test type are shown in Fig. 4. $\Delta\varepsilon_{in}$ was about 1×10^{-3} for the 650C isothermal test and both bithermal tests, but about 5×10^{-3} for the 1050C isothermal test. Dislocation densities are greater for the 650C test and both bithermal tests than for the 1050C test though it had a higher $\Delta\varepsilon_{in}$. However, in both bithermal tests most of the dislocations appear in interfacial networks around the γ' particles, as in the 1050C isothermal test. Only a few dislocations may be observed within the γ' particles for the bithermal tests.

Isothermal Fatigue Life. The isothermal fatigue life of bare and coated PWA 1480 at 650 and 1050C as a function of $\Delta\varepsilon_{in}$ is shown in Fig. 5. The true monotonic tensile ductilities for bare specimens tested at the two temperatures are plotted at 1/4 cycle. Note first the bare specimen lives. Best fit lines have slopes of -.58 and -.50 for 650 and 1050C.

It may be seen in Fig. 5 that for these relatively short life tests there is no effect of the coating on the 1050C fatigue

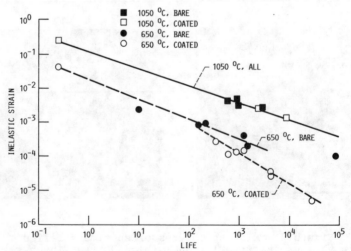

Figure 5. Fatigue life-$\Delta\varepsilon_{in}$ behavior for bare and coated specimens in isothermal tests.

life. The longest test was only about 24h. However, at 650C the coating drastically reduces fatigue life for tests with small $\Delta\varepsilon_{in}$. For $\Delta\varepsilon_{in} \leq 10^{-4}$ coated specimens have a life 1/10th to 1/100th that of bare specimens.

A sectioned coated specimen cycled at 650C to 1/4 of its expected life for an $\Delta\varepsilon_{in}$ of 10^{-4}, 10^3 cycles, shows cracks in the coating already penetrating into the single crystal, Fig. 6(a). A bare single crystal exhibited a life of 10^5 cycles at a slightly higher $\Delta\varepsilon_{in}$. This bare specimen failed at an internal micropore, but the bare specimens tested at higher $\Delta\varepsilon_{in}$ failed at surface initiated cracks. In contrast, at 1050C both coated and bare single crystals failed more frequently by linkage of multiple cracks growing from internal pores, particularly at the lowest strain ranges, Fig 6(b). Though coating cracks did develop, they initiated more slowly than at 650C for the same $\Delta\varepsilon_{in}$ and propagated more slowly in the single crystal.

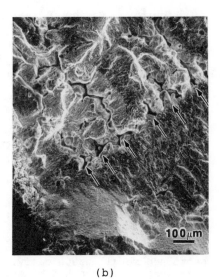

(a)   (b)

Figure 6. Failure origins in coated specimens tested isothermally: (a) Surface cracks in a 650C tests interrupted at 250 cycles (1/4 of life), (b) Cracks initiated at micropores in the superalloy at 1050C.

In-Phase Bithermal Fatigue Life. Life for in-phase bithermal tests of bare and coated specimens on a $\Delta\varepsilon_{in}$ basis is shown relative to the isothermal results in Fig. 7. The in-phase test life for bare specimens appears about the same as the 650C isothermal test life. However, this is somewhat misleading since bare specimens appear to fail as much due to gross oxidation as to fatigue for in-phase tests. A large volume of oxides spalled off the specimens in these tests. Significant oxide spalling did not occur in the 1050C isothermal tests, which does not strain the oxide at low temperature, or in the out-of-phase tests, which were much shorter lived, as will be seen. Further, compressive strains which cause the oxide scale to buckle and spall are small in the out-of-phase test, Fig. 2.

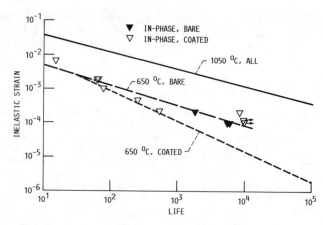

Figure 7. Fatigue life-$\Delta\varepsilon_{in}$ behavior for bare and coated
specimens in in-phase bithermal tests relative to
isothermal fatigue behavior.

Also, in the two load-controlled, lowest-strain in-phase tests
of bare specimens, considerable necking occurred indicating
significantly more ductility than for either 650C isothermal or
out-of-phase tests. Thus, in the low-strain regime, actual
failure due to fatigue in bare specimens in the in-phase test
may be even more difficult than in isothermal 650C cycling.
This view was confirmed by the long lives of coated specimens in
low-strain in-phase tests, Fig. 7. For $\Delta\varepsilon_{in}$ below about 2×10^{-4},
the coating does not rapidly crack as in the 650C isothermal
tests. A specimen cycled at a $\Delta\varepsilon_{in}$ of about 10^{-4} to the number
of cycles that produced failure in the 650C isothermal test
exhibited no coating cracking. Lives of coated specimens in the
in-phase test were at least 10X greater than in the 650C
isothermal test at low strains.

The coated specimens which had not failed in about 10^4 cycles in
the in-phase tests, Fig. 7, were removed from testing and
sectioned metallographically. It may be seen in Fig. 8(a) that
some surface cracks had penetrated the single crystals. However,
these cracks were very blunt relative to secondary cracks
observed in specimens tested isothermally at 650C, Fig. 6(a).
Internal cracks eminating from micropores, also apparent in Fig.
8(a), appeared more likely to lead to failure. A specimen tested
to failure at a slightly higher strain range did fail from
internal micropores, Fig. 8(b).

For the intermediate $\Delta\varepsilon_{in}$ tests shown in Fig. 7, life was
drastically reduced for the in-phase tests of coated specimens
falling near the life line for 650C tests of coated specimens.
This suggests that at these strain ranges as in the 650C tests,
early cracking in the coating may control life. Indeed, only a
few small internal cracks were observed on fracture surfaces,
though it was not clear whether the principle crack initiated in
the coating or at the superalloy surface. Still, it appears that
for the highest strain range tests, the in-phase and 650C test
lives for both bare and coated specimens may all converge. This
would be expected since in the limit all would fail in the first
cycle at a $\Delta\varepsilon_{in}$ equal to the alloy ductility.

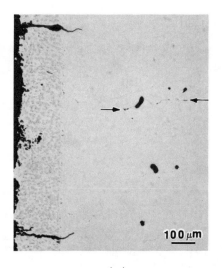 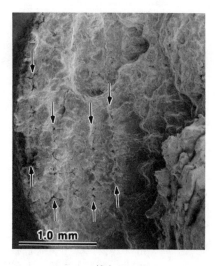

(a)   (b)

Figure 8. Failure origins in coated in-phase test
specimens: (a) Blunt surface cracks and internal
crack in a test interrupted at about 10^4 cycles,
(b) Cracks initiated at micropores in specimen
failed at slightly less than 10^4 cycles.

Out-Of-Phase Bithermal Fatige Life. As shown in Fig. 9, both
coated and bare specimens have short lives in the out-of-phase
tests. Surface cracks initiate early as in the 650C tests of
coated specimens. A test of a coated specimen at a $\Delta\varepsilon_{in}$ of 10^{-4}
interrupted at 1/10th of the expected life exhibited numerous
cracks which had already penetrated into the single crystal.
Fig. 10 shows cracks initiated in the coating in out-of-phase
tests on: (a) a longitudinal section, and (b) a fracture
surface. It may be seen that these cracks appear much sharper
than the cracks shown for in-phase tests in Fig. 8(a). However,
as the results for the bare single crystals show, the coating is
not necessary for rapid crack initiation in the out-of-phase
cycle. The lives of the coated and bare single crystals are
indistinguishable in the strain range regime where
they can be compared.

The detrimental effect of the out-of-phase cycle is actually
much worse than is apparent in Fig. 9. Since many of the tests
were conducted with $\Delta\varepsilon_{in}$ below the resolution limit of about
10^{-4}, lives cannot be compared in that regime on a $\Delta\varepsilon_{in}$ basis.
Fig. 11 shows life on the basis of maximum tensile stress,
σ_{max}, in the cycle for the various bithermal tests and best fit
curves for the isothermal test results. Here the full
detrimental effect of the out-of-phase cycle in the long-life
regime may be seen, and again it may be seen that bare as well
as coated specimens suffer equally in the out-of-phase cycle.

Since crack initiation is observed to occur early in both the
out-of-phase tests and 650C tests of coated specimens, the
shorter life for the out-of-phase tests appears to be due to an

189

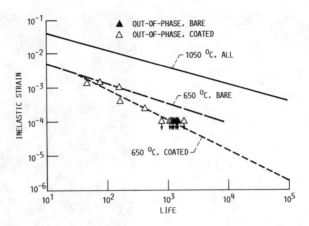

Figure 9. Fatigue life-$\Delta\varepsilon_{in}$ behavior for bare and
coated specimens in out-of-phase bithermal tests
relative to isothermal fatigue behavior.

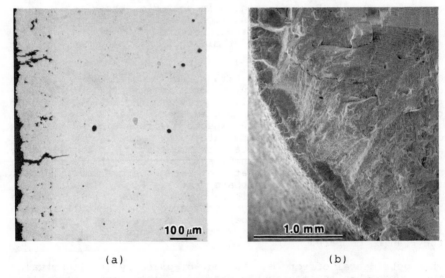

(a)   (b)

Figure 10. Failure origins in coated out-of-phase
bithermal test specimens: (a) Sharp cracks penetrating
into superalloy in a test interrupted at 1/10th of
expected life, (b) Fracture surface showing cracks
initiated in coating.

increased crack growth rate. The detrimental effect of 1050C
compressive half of the out-of-phase cycle appears to be
environmental rather than mechanical. This is suggested by the
results of out-of-phase tests of coated specimens conducted in
vacuum shown in Fig. 12. It may be seen that in the absence of
air the out-of-phase test lives are equal or greater than those
for coated specimens tested at 650C.

190

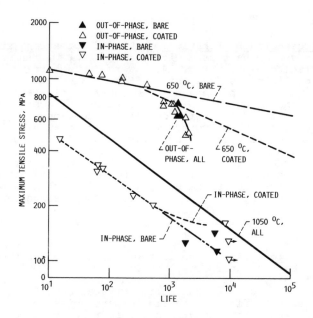

Figure 11. Fatigue life-σ_{max} behavior for bare and coated specimens in all tests.

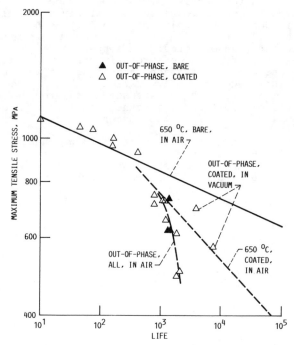

Figure 12. Relative fatigue life-σ_{max} behavior for coated specimens in out-of-phase tests in vacuum.

DISCUSSION

Cyclic Stress-Strain Behavior.   The deformation structures observed in specimens tested with similar $\Delta\varepsilon_{in}$ in the isothermal and bithermal cycles, Fig. 4, were in accordance with the observed cyclic stress responses, Fig. 3.   The 1050C stress responses of the in-phase and out-of-phase cycles are the same, both higher than that for the isothermal cycle. This correlates with the higher dislocation densities developed in the two bithermal cycles. These were, in fact, comparable to those for the 650C isothermal cycle, and correspondingly, the 650C stress response was similar among these three cycle types irrespective of the fact that for the bithermal cycles a greater fraction of the dislocations were in the $\gamma-\gamma'$ interfaces.   The somewhat lower stress for the in-phase cycle, which is in compression, may simply reflect the same tension-compression anisotropy observed in 650C isothermal tests.   The 650C cyclic stress-strain behavior of this alloy is discussed in more detail elsewhere (11).

Isothermal Fatigue Life. The isothermal fatigue behavior of the bare and coated superalloy single crystals, Fig. 5, is straightforward.   The lives of the bare crystals at both temperatures, as well as the coated crystals at 1050C, exhibit classic Manson-Coffin exponential dependencies on $\Delta\varepsilon_{in}$ which extrapolate well to failure at the true monotonic tensile ductility at 1/4 cycle. The slopes of -.50 for the 1050C tests and -.58 for the 650C tests of bare specimens are in the range of typical values for metals.   The difference in life between the two temperatures appears to be simply a result of the difference in ductility. Only the life of the bare crystal at 650C and the lowest $\Delta\varepsilon_{in}$ does not fit this description, appearing longer than expected. This specimen also failed in a different manner, from large internal pore.   At higher $\Delta\varepsilon_{in}$, cracks initiated at the surface and a defect was not always apparent.   This suggests a lower limit in $\Delta\varepsilon_{in}$ necessary for surface crack initiation in the absence of a large defect. Such a transition from surface to internal failure with decreasing strain range is observed for other materials.

More rapid crack initiation in the coating than in the base crystals during 650C fatigue is due to the large $\Delta\varepsilon_{in}$ enforced in the coating by the stronger crystals. Fatigue behavior of the bulk PWA 276 NiCoCrAlY coating is actually superior on a $\Delta\varepsilon_{in}$ basis (8). The $\Delta\varepsilon_{in}$ required to produce failure in 10^4 cycles was 10X greater for the coating than the PWA 1480 superalloy.   However, the coating experiences the same $\Delta\varepsilon_t$ as the underlying crystal, and for small $\Delta\varepsilon_{in}$ which only strain the superalloy elastically, the coating can experience inelastic strain. For a $\Delta\varepsilon_t$ expected to produce failure in the superalloy in 10^4 cycles, the bulk coating had a life of only about 5×10^2. For higher $\Delta\varepsilon_t$, increased $\Delta\varepsilon_{in}$ in the superalloy greatly reduces its life, such that for a $\Delta\varepsilon_t$ of about 2×10^{-2}, both coating and superalloy have lives of about 10^2 cycles. Note in Fig. 5 that this is about the life at which the behavior of the coated and bare specimens begin to merge.

The fatigue study of the bulk NiCoCrAlY alloy showed that it has an enormous tolerance for inelastic strain at 1050C. Thus, even though the coating is very weak relative to the superalloy,

192

Fig. 3(a), and thus suffers much larger $\Delta\varepsilon_{in}$, the life of the bulk coating was about 10X greater than that of PWA 1480 over the range of $\Delta\varepsilon_t$ investigated, $8X10^{-3}$ to $2X10^{-2}$. Thus, failures originating in the coating were not observed in tests at 1050C. And since for the strain ranges employed both the bare and coated specimens failed at internal pores, the same lives were observed. Further, this indicates little environmental damage to the bare specimens, at least in these tests which lasted less than a day.

<u>In-Phase Bithermal Fatigue Life.</u> In the high-strain regime, the in-phase bithermal fatigue lives appear to be controlled by the 650C ductility of the superalloy just as the 650C isothermal fatigue life, Fig. 7. This may be somewhat surprising in that the 650C loading in the in-phase test is compressive, and the meaning of ductility in compression is not clear. However, failure in compression in the single crystal superalloy is not greatly different than in tension. Specimens loaded in compression at 650C will shear into two pieces along a single slip band at few percent strain just as in tension.

In the long-life regime, other mechanisms influence the in-phase test life. For bare specimens, oxidation has an obvious influence, producing considerable loss in load bearing area. Yet, the in-phase test life for bare specimens is not greatly foreshortened relative to the 650C isothermal fatigue life for bare specimens. The very blunt secondary cracks and evidence of considerable ductility, ie. necking, relative to those for specimens tested at 650C appear to indicate that crack growth during the 1050C tensile half of the in-phase cycle is more difficult than at 650C.

Behavior of coated specimens in the intermediate and long-life regime is more complex. In the intermediate life regime the in-phase test life of coated specimens may be controlled by coating cracking as in the 650C isothermal tests. This is suggested by the life similarity, Fig. 7, and fractographic examination. However, it is clear that in the low-strain long-life regime crack initiation in the coating is more difficult in the in-phase test than in the 650C isothermal test. Life increases dramatically for low-$\Delta\varepsilon_{in}$ in-phase tests, and the failure origin changes to cracks initiated at internal micropores. Further study of this apparently abrupt change in behavior for the in-phase test is required.

<u>Out-Of-Phase Bithermal Fatigue Life.</u> The surprise for out-of-phase tests is that lives of both bare and coated specimens are foreshortened by rapid initiation of surface cracks. Though for coated specimens, crack initiation may be due to the large enforced tensile strain in the coating at 650C just as for the 650C isothermal tests, rapid surface cracking also occurs in bare specimens. The similarity in life for coated and bare specimens suggests that crack initiation represents an insignificant fraction of life and that crack propagation in the base superalloy is the measure of life.

Since crack initiation is also rapid for 650C isothermal tests of coated specimens, the shorter life for the out-of-phase tests is probably due to more rapid crack propagation. That this is an environmental effect of the 1050C exposure is suggested by the increase in life for out-of-phase tests of coated specimens conducted in vacuum. These exhibited lives as long as for the 650C isothermal tests of coated specimens. This finding and the rapid crack initiation in bare specimens for the out-of-phase test suggest a mechanism whereby a surface layer of oxide or dissolved oxygen forms during the 1050C half of the cycle and subsequently cracks brittlely during the 650C tensile loading. To be consistent one would have to postulate that this layer has more ductility at 1050C, and thus does not lead to rapid crack initiation or propagation in the cycles imposing tensile loading at 1050C, the isothermal or in-phase tests.

While the lives of bare and coated specimens are the same in out-of-phase tests in air it would be misleading to conclude that the coating has no effect on life in the out-of-phase test. It is reasonable to assume that if the coating did not fail, the superalloy beneath, protected from the environment, would have a much longer life. Even for the out-of-phase tests in vacuum rapid crack initiation probably occurs in the coating. In the 650C isothermal tests of bare specimens, where rapid surface crack initiation does not occur, lives are much longer than for the out-of-phase tests in vacuum. For the same reason, it is expected that out-of-phase test lives of bare specimens in vacuum would be much longer than those of coated specimens.

It may be somewhat surprising, that lives for the coated specimens in the out-of-phase tests in vacuum are not shorter than for those in the 650C isothermal tests. For the out-of-phase test, the tensile thermal mismatch strain in the coating upon cooling makes $\Delta\varepsilon_t$ greater than the applied tensile strain. The coefficient of thermal expansion in the coating averages 1.8×10^{-6} C^{-1} greater than that of the superalloy over the temperature range of these tests (11). So, at most, the thermal strain in the coating upon cooling from 1050 to 650C could be about 7×10^{-4}, which would certainly be very damaging.

Two probable explanations come to mind. First, the coating is so weak at the higher temperatures in the cycle (6) that some of the developing thermal mismatch stress probably creeps out during cooling. Thus, less than the full thermal mismatch strain may remain as inelastic strain in the coating at the start of the 650C half of the cycle. Second, the first coating cracks develops so rapidly just under the influence of the applied strain ranges in the 650C isothermal tests conducted herein, that any acceleration due to the added thermal mismatch strain in the out-of-phase cycles is unnoticable.

It should be recognized that the bithermal out-of-phase test is not only an idealized, but also a very severe test relative to what might be experienced in a gas turbine engine component because all the tensile strain is imposed in the temperature range where the materials have least ductility. The goal must be to eventually predict when component design and engine operating conditions will lead to premature failure in the coating rather than its intended function of environmental protection.

Further work is needed to confirm that rapid crack propagation and early crack initiation in bare specimens is largely an environmental effect in the out-of-phase test. There is,

194

however, evidence that mechanical effects of the 1050C exposure, say creep damage, are not large. It has been shown that the 1050C creep-fatigue behavior of PWA 1480 can be explained by the $\Delta\varepsilon_{in}$ and the stress range without regard to the proportions of creep and rapid plastic strain (4). Also, the cyclic crack propagation rate of PWA 1480 has been shown to be insensitive to cycle frequency by DeLuca and Cowles (9), albeit at 982C. The isothermal and bithermal fatigue behavior of the bare specimens observed herein can be rationalized based on the ductility of PWA 1480 at the two temperatures and a proposed environmental effect in the out-of-phase cycle.

It should be pointed out that cyclic crack propagation rates were found by DeLuca and Cowles to be the same for tests of PWA 1480 in more conventional out-of-phase and in-phase TMF cycles between 982 and 427C. These tests employed stress ratios approximating those developed in reversed strain fatigue tests such as those conducted herein. However, because of constraints on testing time, only crack growth rates greater than 10^{-5} mm/cycle were studied. It is reasonable that the environmental effect on the out-of-phase crack propagation suggested by the present study appears only in the near threshold regime where the crack advance per cycle is on the order of the depth of material which could be affected by oxidation during the 1050C half of the cycle. The bulk of crack propagation life in fatigue tests is spent in this regime, and thus any effect on propagation rate would be most noticable.

Thermomechanical Fatigue Life. Finally, we should comment on the suitability of the simplified, bithermal test in understanding TMF behavior. The bithermal test shows the basic connection between TMF and isothermal fatigue behavior on a $\Delta\varepsilon_{in}$ basis in the high-strain regime where the basic mechanical damage processes in the superalloy control life. High-strain life is controlled by the ductility of the superalloy, and for the single crystal material this corresponds to the low temperature half of the bithermal cycle even when the loading is compressive. Further, this view of the basic TMF behavior of the superalloy provides a norm against which to judge the effects of the other damage mechanisms operating in the low-strain long-life regime, coating induced cracking and the environment. It should be pointed out however, that the simple correspondence between the low temperature isothermal and both bithermal cycles in the high-strain regime should not be expected in all materials. At the low temperature the single crystal superalloy has low ductility in compression as well as in tension, and the material is not greatly changed in the bithermal cycles relative to the microstructure developed during 650C isothermal cycling. This would not be the same for polycrystalline materials exhibiting grain boundary cavitation or alloys with less stable precipitates.

For coated specimens in the long life regime it is also necessary to consider the strain-temperature cycle in the coating since cracking in the coating can reduce the cycles necessary to develop a crack in the superalloy and exposure to the atmosphere may accelerate crack growth as shown for the out-of-phase cycle. Reasonable predictions of the cycles necessary to crack the coating in the 650C isothermal fatigue were obtained by estimating the $\Delta\varepsilon_{in}$ in the coating as shown in

the Appendix and using the fatigue life data for the bulk
coating alloy (8). However, it was necessary to use only about
10 o/o of the life to total separation, which corresponded to
noticible surface cracking in the bulk coating specimens. Such
an approach to the more important task of predicting TMF life
based on $\Delta\varepsilon_{in}$ in the coating will require more understanding of
the constitutive behavior of the coating. Further, life
prediction relevant to actual gas turbine engine components will
be further complicated by the need to account for other damage
mechanisms occuring in long time service, such as gross
oxidation or hot corrosion, and interdiffusion between coating
and superalloy.

For more complicated TMF cycles it is common to compare life
based on $\Delta\varepsilon_t$ because of the difficulty in separating out $\Delta\varepsilon_{in}$.
Success has been shown in using the applied $\Delta\varepsilon_t$ plus the thermal
mismatch strain to correlate coating crack initiation for
various coating-alloy combinations with different mismatches in
the coefficient of thermal expansion (2,3). However, for the
very different types of fatigue cycles studied herein,
consideration of $\Delta\varepsilon_t$ alone does not contribute much to
understanding. The lives for coated specimens in the two
isothermal and two bithermal tests are shown on the basis of
$\Delta\varepsilon_t$ in Fig. 13. It is difficult to see the connection between
the two bithermal tests, the 650C isothermal test, and the 650C
ductility demonstrated by considering the $\Delta\varepsilon_{in}$.

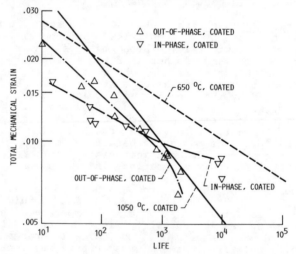

Figure 13. Fatigue life-$\Delta\varepsilon_t$ behavior for coated
specimens in all tests.

RESULTS AND CONCLUSIONS

Specimens of a single crystal superalloy, PWA 1480, both bare
and coated with a NiCoCrAlY alloy, PWA 276, were tested in
fatigue at 650 and 1050C and in TMF tests between the two
temperatures. A 'bithermal' TMF test type was employed in which
mechanical loading was applied at the temperature endpoints.
The test imposing tensile and compressive loading at the high

and low temperatures, respectively, is termed 'in-phase', and the opposite cycle is termed 'out-of-phase'. The following results and conclusions were obtained.

1. The only significant difference in cyclic stress response among the test types was a lower stress at 1050C for the isothermal test than for the bithermal tests. This correlated with an observed lower dislocation density than for the tests involving deformation at 650C.
2. Isothermal fatigue life on a $\Delta\varepsilon_{in}$ basis was as expected based on the monotonic tensile ductility of the superalloy. Tensile ductility and fatigue life are both much lower at 650 than at 1050C.
3. For high $\Delta\varepsilon_{in}$, lives for both in- and out-of-phase bithermal tests approached that for the 650C isothermal tests. All are limited by the 650C ductility of the superalloy.
4. For low $\Delta\varepsilon_{in}$, the coating reduced life in 650C isothermal tests. Cracks initiated early in the coating and propagated into the superalloy.
5. Cracks initiated rapidly in bare as well as coated specimens in the out-of-phase tests and these propagated faster than in the 650C isothermal test. Tests in vacuum suggested this is an environmental effect due to the 1050C exposure and subsequent tensile straining at 650C.
6. For low $\Delta\varepsilon_{in}$, the coating improved life for in-phase tests.
7. The simplified bithermal TMF tests exhibit the same effects effects of more realistic TMF tests and show the connection with isothermal fatigue behavior in the high-$\Delta\varepsilon_{in}$ regime where the mechanical damage mechanisms in the superalloy control life. This view of basic TMF of the superalloy also provides a norm against which the effects of the other damage mechanisms operating in the low-$\Delta\varepsilon_{in}$, long-life regime, coating initiated cracking and the environment, may be assessed.

REFERENCES

1. K. Schneider and H. W. Grunling, Thin Solid Films, 107(1983), p. 395.
2. G. R. Leverant, T. E. Strangman, and B. S. Langer, Superalloys: Metallurgy and Manufacture, Claitor's, Baton Rouge, Louisiana, 1976, p.285.
3. T. E. Strangman, Thin Solid Films, 45(1977), p. 499.
4. M. G. Hebsur, and R. V. Miner, NASA TM-88950, National Aeronautics and Space Administration, 1987.
5. R. V. Miner, J. Gayda, and M. G. Hebsur, Low Cycle Fatigue - Directions for the Future, ASTM STP-942, American Society for Testing and Materials, 1987, (in press).
6. M. G. Hebsur, and R. V. Miner, Mat. Sci. Eng., 83(1986), p. 239.
7. M. G. Hebsur, and R. V. Miner, Thin Solid Films, to be published.
8. J. Gayda, T. P. Gabb, and R. V. Miner, Int. J. Fatigue, 8(4)(1986), p. 217.
9. D. P. DeLuca, and B. A. Cowles, AFWAL-TR-84-4167, Air Force Aeronautical Laboratories, Wright-Patterson AFB, Ohio, 1984.
10. G. R. Halford, M. A. McGaw, R. C. Bill, and P. D. Fanti, Low Cycle Fatigue - Directions for the Future, ASTM

STP-942, American Society for Testing and Materials, 1987, (in press).
11. T. P. Gabb, and G. E. Welsch, Scripta Met., **20**(1986), p. 1049.
12. B. A. Cowles, private communication, Pratt and Whitney Aircraft, West Palm Beach, Florida.

APPENDIX

In the 650C isothermal test and the 650C halves of the bithermal tests, the coating on the relatively small fatigue specimens used herein carries a non-negligable load. This Appendix describes estimation of the actual stresses and strains in the coating and superalloy in the 650C isothermal tests. Since the coating carries very little load at 1050C, the actual strains in the superalloy are what is measured in the 1050C isothermal tests and the 1050C halves of the bithermal tests.

The elastic and inelastic strains in the superalloy and coating were calculated by requiring continuity and balance of forces. The elastic strains were calculated based on elastic moduli of 115 and 135 GPa, respectively for superalloy and coating at 650C. The inelastic strains were based on the cyclic stress-strain curves developed herein for the bare superalloy specimens and previously for the bulk coating (8). In an equation of the form

$$\sigma_{max} = a\Delta\varepsilon_{in}^{n'},$$

where σ_{max} is the maximum stress, a and n' were taken as 2900 and .15 for the superalloy, and 1350 and .193 for the coating. Equating the total strain in the superalloy and coating, and requiring the load on the superalloy to equal the applied load minus the load on the coating, the load on the coating is obtained for a given applied load. With these estimates of the true loads in the coating and superalloy, the corresponding estimates of the inelastic strains were obtained from the above cyclic stress-strain equations.

Since the coating experiences a greater inelastic strain than the superalloy, upon removal of, say, a tensile load the requirement of continuity drives the coating into compression and the superalloy is held in slight tension. By again requiring a balance of forces, the small elastic tensile strain in the superalloy at zero load may be calculated. This small elastic strain appears in addition to the true inelastic strain at zero load, making the apparent inelastic strain too large.

Microstructure, Fracture and Damage

INFLUENCE OF THERMAL AGING ON THE CREEP CRACK GROWTH BEHAVIOR

OF A Cr-Mo STEEL

S. Jani and A. Saxena

School of Materials Engineering
Georgia Institute of Technology
Atlanta, GA 30332-0245

Abstract

More realistic analyses of the remaining life of high temperature components are achieved by utilizing the recently developed approach of time-dependent fracture mechanics (TDFM) in conjunction with the conventional creep deformation and rupture approaches. It is shown that in-service thermal aging or degradation can adversely influence a component's creep deformation and rupture behavior. Further, it is shown that the in-service degradation is due to grain boundary (GB) carbide coarsening. It is argued that the coarsened GB carbide structure only mildly influences the kinetics of GB creep cavitation, which is the process by which creep crack growth occurs. Thus, a strong influence of in-service degradation on the creep crack growth rate expressed as a function of the C_t parameter is not expected nor was it observed.

Introduction

The need within the power generation industry to estimate remaining lives of high temperature structural components which are approaching or have passed their original design life has been strong due to considerations of safety and the cost of forced outages. These considerations provide an impetus for instituting a philosophy of "retirement for a cause" and for developing a methodology for predicting the remaining life in high temperature components. Currently used methods of life-fraction rules and accelerated creep rupture tests are limited due to their dependence on extrapolation of inherently scattered data. Furthermore, the application of uniaxial creep deformation and rupture data for predicting the service life of a thick-section component is questionable because these components experience considerable temperature and stress gradients. As a consequence, they usually fail by the nucleation and/or growth or localized damage, specifically cracks. Figure 1 (courtesy of Babcock and Wilcox Co.) shows the problem of cracking between ligaments of tube holes in a secondary superheater outlet header which had been in service at Mississippi Power and Light for twenty four years.

Figure 1   Inside view of extensive ligament cracking in a secondary stage superheater outlet header (Courtesy of Babcock and Wilcox, Co.)

Since crack nucleation and propagation is a localized phenomenon, fracture mechanics type analysis is more appropriate for life prediction. However, the well established concepts of linear elastic fracture mechanics (LEFM) and elastic-plastic fracture mechanics (EPFM) cannot be applied for correlating creep crack growth in creep ductile materials such as Cr-Mo steels and austenitic stainless steels. Time-dependent fracture mechanics (TDFM) concepts have recently been developed to tackle this problem and will be described briefly in the next section.

In predicting service life of high temperature components, there is an added complication associated with microstructural transformations, referred to as aging or degradation, during service. This in-service aging can adversely affect creep and fatigue properties and therefore a methodology for estimating remaining life cannot be based solely on virgin-material properties. Rather, the effect of thermal aging on mechanical properties has to be incorporated into such a methodology.

The present study was conducted to investigate the effect of in-service thermal aging on the creep and crack growth behavior of a 1¼Cr-½Mo steel used in power plant components such as steam headers and large pipes.

Time-Dependent Fracture Mechanics

The concepts of time-dependent fracture mechanics (TDFM) provide a parameter, C_t, for correlating creep crack growth data in creep-ductile materials. To be considered valid, such a parameter should be able to uniquely describe the crack tip conditions. The basis of the C_t parameter is briefly described in this section.

When a cracked body is loaded at a high temperature where creep deformation occurs, then the deformation scenario ahead of the crack tip will be as shown in Figure 2. On loading, a plastic zone forms immediately, the size of which will depend on the applied stress intensity parameter, K and the yield strength of the material. Also shown in the figure is the K-zone size within which the elastic stress and deformation fields apply. The J-integral characterizes the fields if the plastic zone becomes comparable to the K-zone in size. With time, the stresses in the vicinity of the crack will begin to relax due to creep deformation. The creep zone will thereafter grow with time. Now K and the J-integral by themselves are not valid crack tip parameters. Therefore, new parameters are required to characterize the stress and deformation fields and the crack growth under creep conditions. This forms the basis for TDFM.

The scale of creep deformation in a cracked body is an important consideration in TDFM (analogous to the extent of plastic deformation in LEFM and EPFM). The regimes of creep deformation in a cracked body are shown schematically in Figure 3:

1. Small Scale Creep (SSC): The creep zone is small compared to the crack length and the remaining ligament of the body.
2. Transition Creep (TC): The creep zone has grown to a size comparable to the pertinent dimensions of the body.
3. Steady State (SS) or Extensive Creep: Here the entire remaining ligament has been engulfed by creep and the crack tip stresses no longer change with time.

SSC and TC regimes are highly transient conditions during which crack tip stresses are changing with time.

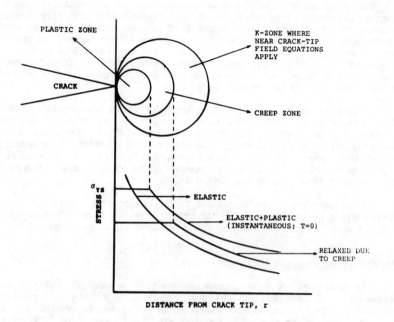

Figure 2   Schematic representation of the deformation zones ahead of a crack tip loaded at an elevated temperature.

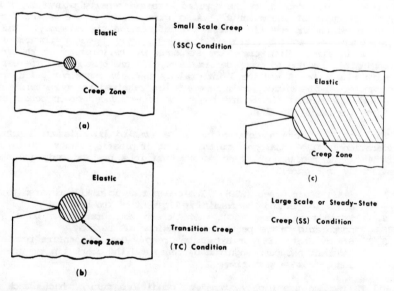

Figure 3   Schematic representation of the extent of creep deformation under which creep-crack growth can occur.

Under SS, creep crack growth has been characterized by the path-independent C^*-integral [1-3], provided that power law creep applies:

$$C^* = \int_\Gamma W^* \, dy - T_i \frac{d\dot{u}_i}{dx} \, d_s \tag{1}$$

where W^* is the strain-rate energy density and is given by:

$$W^* = \int_0^{\dot{\varepsilon}_{ij}} \sigma_{ij} \cdot d\dot{\varepsilon}_{ij} \tag{2}$$

T_i is the traction vector along the path Γ (considering the region enclosed by the path as a free body) which originates at a point along the lower crack surface, goes counter-clockwise, and terminates at a point on the upper crack surface. $\dot{u}_i$ is the deflection rate vector along the direction of traction. σ_{ij} and $\dot{\varepsilon}_{ij}$ are the stress and strain rate tensors respectively. C^* has also been interpreted as the difference between the energy rates (or power) supplied to two bodies with incrementally differing crack lengths [1].

The expression for determining C^* is expressed in the following form [4]:

$$C^* = \frac{P\dot{V}_c}{BW} \eta(a/W) \tag{3}$$

where P is the load, $\dot{V}_c$ is the steady state load line deflection rate due to creep, B is the thickness and W is the width of the specimen. For compact type (CT) specimens,

$$\eta(a/W) = \frac{1}{(1-a/W)} \cdot \frac{n}{(n+1)} \cdot (\gamma-\beta/n) \tag{4}$$

where a is the crack length, α and β are constants listed as a function of (a/W) and n [5,6].

The following equations relate C^* to the crack tip stress and strain rate [7]:

$$\sigma_{ij} = \frac{C^*}{AI_n}^{\frac{1}{n+1}} \cdot r^{-\frac{1}{n+1}} \cdot \tilde{\sigma}(\theta) \tag{5a}$$

$$\dot{\varepsilon}_{ij} = \frac{C^*}{AI_n}^{\frac{n}{n+1}} \cdot r^{-\frac{n}{n+1}} \cdot \dot{\tilde{\varepsilon}}(\theta) \tag{5b}$$

where r is the distance from the crack tip, θ is the angle from the plane of the crack, $\sigma(\theta)$ and $\tilde{\varepsilon}(\theta)$ are angular functions listed in Reference [8], I_n is a constant dependent on the steady state creep exponent (n) and is also listed [8], and, A and n are power-law creep coefficient and exponent respectively.

Thus c^* characterizes the strength of the crack tip stress singularity commonly known as the Hutchinson, Rice and Rosengren (HRR) singularity [9,10] and therefore is expected to characterize creep crack growths. This has been demonstrated by several experimental studies [1-3].

The validity of the c^*-integral is limited to the SS regime. However, in practice, many high temperature components are thick-section and SS conditions may not be present in all cases. The parameter proposed to characterize the behavior in the entire region of SSC through SS is the C_t parameter [11]. The C_t parameter is based on a generalization of the energy-rate interpretation of c^* as follows: Figure 4 shows a schematic of the behavior of several identical pairs of cracked specimens with crack lengths a and a + Δa respectively. These pairs are loaded to various levels, P_1, P_2, P_3, . . . P_i, at an elevated temperature. V_c is the load line deflection due to creep and $\dot{V}_c$ is its corresponding time-rate. Then C_t is given by:

$$C_t = \lim_{\Delta a \to 0} \left(-\frac{1}{B} \frac{\Delta U_t^*}{\Delta a} \right) = -\frac{1}{B} \frac{\partial U_t^*}{\partial a} \qquad (6)$$

where the area, ΔU_t^* represents the difference in the energy rates (or power) supplied to the two cracked bodies, B is the specimen thickness. At infinitely long times, $C_t = c^*$.

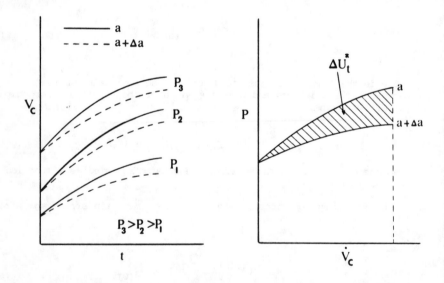

Figure 4   Schematic of the energy-rate interpretation of the C_t parameter

It has been shown that the C_t parameter correlates creep crack growth well in Cr-Mo steels over a wide range or creep crack growth rates [11,12]. Recent work [13,14] has shown that in the SSC regime, C_t uniquely characterizes the rate of increase of the crack tip creep zone size. It was also shown that in the SSC regime, C_t does not characterize the instantaneous crack tip stress and strain rate fields as does C^* in the SS regime (Equation 5). However, through its relationship with the creep zone size, perhaps C_t accurately captures the history-dependence of the creep damage accumulation at the crack tip and thus, correlates with the crack growth rates under transient conditions. In any event, there is considerable evidence of correlation between creep crack growth rate and C_t over a wide variety of conditions [15]. The method for estimating C_t in test specimens is described in the next section.

Experimental

Test Material

The test material was taken from a secondary stage superheater outlet header that was retired from service after 24 years due to ligament cracking (see Figure 1). The chemical composition of the material is given in Table 1. The extent of service aging, or "damage" was expected to vary along the 55 foot length of the header due to a service temperature gradient. the test material was extracted from the two extreme ends of the header labelled the north-end and the south-end, also called the hot-end and the cold-end, respectively. the measured service temperatures of the North and South ends were 538 C (1000 F) and 510 C (950 F), respectively.

Table I

Chemical Composition of the Test Material (in Weight Percent)

C	Mn	P	S	Si	Cr	Mo
0.1	0.45	0.013	0.013	0.63	1.19	0.61

Tensile Testing

Uniaxial tensile tests were conducted on the cold-end at 24 C, 482 C and 538 C (75, 900 and 1000 F). All tensile testing was conducted using standard 50.8 mm (2 inch) gage length specimens. The stress-strain curves were fitted to the following equation:

$$\varepsilon = \frac{\sigma}{E} + D\sigma^m \qquad (7)$$

where E is the Young's modulus, and D and m are the plasticity coefficient and exponent respectively. These values are required for the calculation of the crack tip parameter, C_t.

Creep Deformation and Rupture Tests

Standard tensile creep specimens with a diameter of 12.83 mm (0.505 inch) and a gage length of 76.2 mm (3 inches) were used. Creep tests were conducted at 538 C (1000 F) in accordance with ASTM Test Standard E139-79 [16] using lever type dead-weight loaded machines. The applied stress levels used were 48.2, 68.9, 82.7, 103.3 and 137.8 MPa (7, 10, 12, 15 and 20 Ksi). A strain versus time curve was obtained for each test and from this, the secondary stage (steady-state or minimum) creep rate, $\dot{\varepsilon}_{ss}$, as a function of applied stress, σ, was obtained and fitted to the power-law creep equation,

$$\dot{\varepsilon}_{ss} = A\sigma^n \tag{8}$$

Creep Crack Growth Testing

All crack growth studies were conducted at 538 C (1000 F). Standard 25.4 mm (1 inch) thick compact type (1T-CT) specimens were used for creep crack growth rate (CCGR) tests. Tests were carried out on lever type dead-weight loaded machines. The load-line deflection and crack length were monitored as a function of time throughout each test. The DC electric potential drop technique [17-19] was used to monitor the crack length. The magnitude of C_t was calculated from the energy-rate interpretation of C_t as follows:

$$C_t = \frac{P\dot{V}_c}{BW} \frac{F'}{F} - C^* \left[\frac{F'/F}{\eta} - 1 \right] \tag{9}$$

where V_c is the load-line deflection rate due to creep deformation, B is the thickness and W is width of the specimen. The remaining terms have been defined earlier. The derivation of Equation (9) does not assume any specific material constitutive behavior and is therefore general. For example, it includes contributions from primary creep to the value of C_t. Readers are referred to references 11 and 12 for details of the derivation of methods for calculation C_t.

Results and Discussion

Creep Deformation and Rupture Behavior

Results of the tensile tests of the cold end material are summarized in Table 2. The tensile properties of the two ends were essentially comparable.

Results of creep deformation and rupture properties are shown in Table 3. At each applied stress level, the hot end exhibited higher steady state creep rate ($\dot{\varepsilon}_{ss}$), lower rupture time (t_r) and higher rupture strain (ε_r) compared to the cold end.

Table II

Tensile Properties of the Cold End Material

Test Temperature °F	°C	E x 10³ ksi	MPa	0.2% Yield Strength ksi	MPa	Ultimate Tensile Strength ksi	MPa	El %	RA %	D ksi⁻ᵐ	MPa⁻ᵐ	m
74	24	30.0	206.8	31.05	214.0	63.0	434.3	36.2	73.1	5.67×10^{-10}	9.56×10^{-14}	4.5
900	482	29.3	202	19.6	135.1	40.6	279.9	39.8	78.8	4.15×10^{-10}	2.20×10^{-14}	5.1
1000	538	20.4	140.6	19.0	131.0	32.35	223.0	49.0	84.8	2.82×10^{-10}	8.36×10^{-15}	5.4

Table III

Creep Deformation and Rupture Properties of the Test Material

Location	Applied Stress MPa	ksi	Steady-State Creep Rate hr^{-1}	Rupture Time hr	Rupture Strain %
North End (hot)	48.2	7	1.2×10^{-6}	6089[*]	0.93[*]
South End (cold)	48.2	7	1.2×10^{-7}	5802[*]	0.092[*]
North End (hot)	68.9	10	2.4×10^{-5}	>7750	>21.4
South End (cold)	68.9	10	9.0×10^{-6}	>7852	>6.33
North End (hot)	82.7	12	9.3×10^{-5}	2755	73.8
South End (cold)	82.7	12	4.2×10^{-5}	5557	70.8
North End (hot)	103.3	15	4.6×10^{-4}	527	73
South End (cold)	103.3	15	3.2×10^{-4}	814	64.3
North End (hot)	137.8	20	7.1×10^{-3}	33.5	54
South End (cold)	137.8	20	5.5×10^{-3}	44	52.1

[*]Tests were terminated.

The steady state creep rate, $\dot{\varepsilon}_{ss}$, and the constants n and A in the power-law creep equation (Eq. 11) are essential for calculation of C_t. A plot of $\dot{\varepsilon}_{ss}$ versus the applied stress is shown in Figure 5 for the two ends. A difference in $\dot{\varepsilon}_{ss}$ of an order of magnitude is seen between the two materials at a stress level of 48.2 MPa (7 Ksi). This difference decreases with increasing stress level until at 137.8 MPa (20 Ksi) the creep rates for the two ends are essentially equal. The constants n and A were obtained from regression analysis of the data and are listed in Table 4.

Table IV

Secondary-Stage Creep Rate Constants for the Test Material at 538°C(1000°F)

Location	n	A $(MPa)^{-n} \cdot hr^{-1}$	$(ksi)^{-n} \cdot hr^{-1}$
North End (Hot)	8.0	4.49×10^{-20}	2.29×10^{-13}
South End (Cold)	10.1	1.462×10^{-24}	4.3×10^{-16}

Rupture time data are plotted against applied stress in Figure 6. At each stress level, the rupture time of the hot end material is shorter than that of the cold end material. This behavior is consistent with the creep deformation behavior. Also shown in Figure 6 is the ASME scatter band [20] for this grade of material, showing that the header material had undergone in-service degradation, with the hot end being more severely degraded.

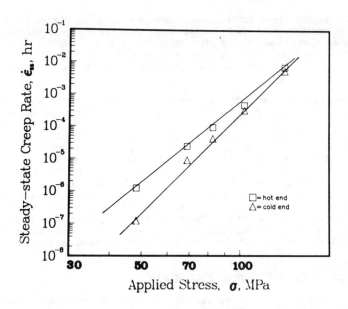

Figure 5   Steady–state creep rate as a function of applied stress for the hot and cold end materials.

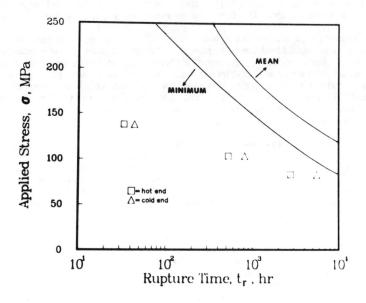

Figure 6   Stress versus rupture time for the hot and cold end materials.

Creep Crack Growth Behavior

Results of the creep crack growth tests are shown in Figure 7. The hot end material exhibits a slightly higher da/dt Vs C_t behavior in the low crack growth regime (less than 2.5 x 10^{-3} mm/hr or 1 x 10^{-4} in/hr). At crack growth rates higher than 5 x 10^{-3} mm/hr or 2 x 10^{-4} in/hr, the creep crack growth rate (CCGR) behavior of the specimens from the two ends is almost identical.

The results of the CCGR experiments indicate that the C_t parameter normalizes to some extent the influence of material state (hot versus cold end materials). However, this does not imply that under identical loading conditions the two ends will exhibit equivalent creep crack growth lives. The reason for this is that the hot end undergoes higher deformation rates and hence higher creep load-line deflection rates. Since the evaluation of C_t involves a product of load and load-line deflection rate due to creep, for identically loaded specimens of the same crack length, the C_t value will be higher for the hot end, and consequently so will the creep crack growth rates. Creep crack growth life of the hot end will therefore be shorter than that of the cold end. Figure 8 shows the remaining life of a header as a function of initial crack size [21]. It is seen that the remaining life is significantly shorter when the hot end properties are used in its evaluation.

Micromechanistic Model for Creep Crack Growth

It is evident from the results of the creep deformation, rupture and crack growth behavior that the hot end material is degraded more severely than cold end material. This is most likely due to the temperature difference (27 C) between the two ends of the header, experienced over an extended period of service. A metallographic analysis was undertaken to characterize the differences between the material states of the two ends. Although this study is not fully complete, the results available at the present time do shed some light on the phenomenon, albeit only qualitatively.

No significant differences were observed in the microstructure of the two ends at low magnifications, Figure 9. The grain size of both ends ranged between ASTM number 6 to 7. However, differences in the microstructure were delineated at higher magnifications as can be seen in Figure 10. It was seen that the grain boundary (GB) carbides in the hot end were coarser. One important observation was that there was no creep cavitation in either material even when observed under thin foil transmission electron microscopy (TEM), Figure 11.

The differences observed in the creep properties are most likely due to thermal softening as discussed by previous researchers [22]. There is a further possibility of other effects, notably the segregation of embrittling species to the grain boundaries [23]. However, no trend toward embrittlement could be detected from the rupture ductilities. Hence, thermal aging appears to be the only rationale for the observed differences in the creep behavior between the hot and the cold ends.

A model based on growth and coalescence of GB creep cavities under the action of crack tip stress fields has shown promise toward explaining some of the observed trends and is discussed in some detail. Creep rupture and creep crack growth occurs dominantly by intergranular cavitation [24]. It is seen in Figure 12 that cracking in Cr-Mo steels does follow an intergranular path and that there is GB cavitation ahead of the crack tip. Furthermore, cavitation ahead of the crack tip was isolated to only a few grain boundaries, suggesting a constrained cavitation mechanism for crack growth and rupture [24].

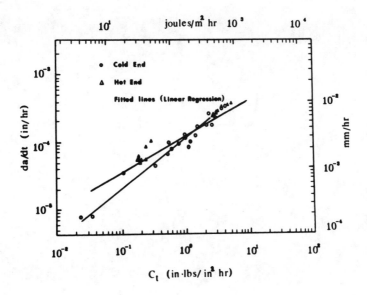

Figure 7   Creep crack growth behavior of the hot and cold end materials.

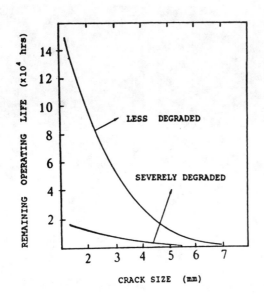

Figure 8   Remaining operating life as a function of initial crack size for two levels of degradation.

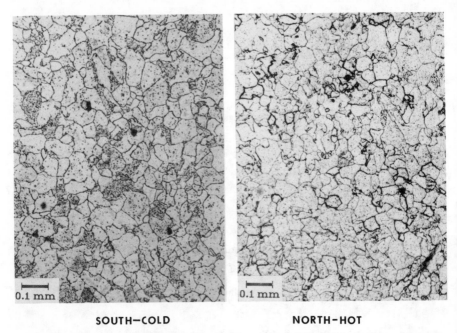

SOUTH—COLD   NORTH-HOT

Figure 9   Low magnification optical micrographs of the hot and cold end materials (Nital etch).

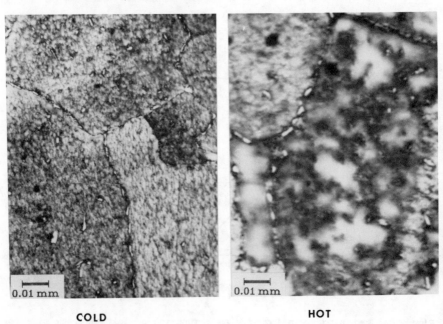

COLD   HOT

Figure 10   High magnification micrographs of the hot and cold end materials (Nital etch).

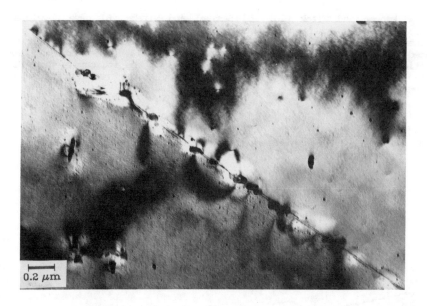

Figure 11   TEM micrograph of the cold end material .

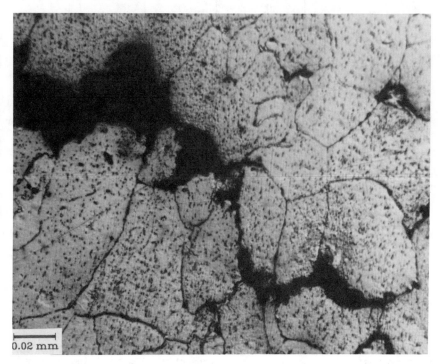

Figure 12   Optical micrograph of the crack tip region of an interrupted cold
end CCGR test specimen.

Wilkinson and Vitek [25] and Bassani and Vitek [26] have developed models for creep crack growth by growth and coalescence of cavities under the influence of crack tip stress fields. They considered an array of N cavities spaced 2b apart, ahead of the crack tip, Figure 13. The radii of the cavities are represented by ρ_i, where i represents the ith cavity from the crack tip. The cavities grow under the influence of the crack tip stress as follows:

$$\dot{\rho} = \phi(\rho,b)\sigma^{\alpha} \qquad (10)$$

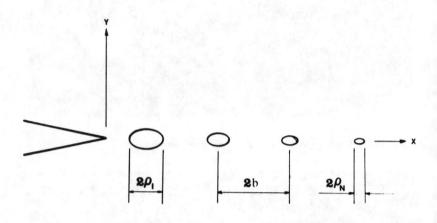

Figure 13 Schematic of an array of grain boundary cavities spaced 2b apart growing under the action of crack tip stress field.

where, ρ is the time rate of change in radius of cavities, and α and the function ϕ depend on the mechanism of cavity growth. The crack tip stress field under steady state (SS) creep is characterized by the HRR field as:

$$\sigma(r,\theta) = \left[\frac{C(t)}{I_n A r}\right]^{\frac{1}{n+1}} \tilde{\sigma}(\theta) \qquad (11)$$

where C(t) is the amplitude of the HRR singularity and equals C^* for SS creep conditions. The other terms in the Eq. 14 have already been defined.

For simplicity, only a one dimensional array of cavities were considered and angular dependence of the stress fields was dropped since only qualitative trends are being considered presently. The criterion for cavity coalescence is that the crack advances by a distance 2b when the cavity nearest to the crack tip grows to a critical size, ρ_c. If this occurs over a

216

time interval Δt, the crack growth rate is

$$\frac{da}{dt} = \lim_{\Delta t \to 0} \frac{2b}{\Delta t} \tag{12}$$

During the time interval, Δt, the cavity with radius ρ_2 grows to a radius ρ_1 and a new cavity of radius ρ_N nucleates at a distance $2bN$ from the crack tip. In this manner, a steady state crack growth process can be established which leads to the following set of N integral equations:

$$\Delta t = \int_{\rho_m}^{\rho_{m-1}} \frac{d\rho}{\dot\rho} \qquad m = 0,1,2, \ldots N-1 \tag{13}$$

where $\rho_0 = \rho_c$. The solution of these N integral equations yields

$$\frac{da}{dt} = \frac{(2b)^{\frac{1-\alpha}{n+1}}}{\psi_0 - \psi_N} \frac{1}{\sum\limits_{m=1}^{N} (m)^{\frac{\alpha}{n+1}}} \left(\frac{C^*}{I_n A}\right)^{\alpha/n+1} \tag{14}$$

where

$$\psi = \int \frac{d\rho}{\phi(\rho,b)} \tag{15}$$

For constrained cavity growth, $\alpha = n$ [27], and

$$\phi = \frac{1}{2.5} \left(\frac{2b}{\rho}\right)^2 d \cdot A \tag{16}$$

where d is the grain diameter and A is the power law creep coefficient. Therefore

$$\psi = \frac{d\rho}{\phi(\rho,b)} = \frac{2.5\rho^3}{3(2b)^2} d \cdot A \tag{17}$$

and

$$\psi_0 - \psi_N = \frac{2.5}{3(2b)^2 d \cdot A} (\rho_c^3 - \rho_N^3) \tag{18}$$

Substituting Equation 18 into 14

$$\frac{da}{dt} = \frac{(2b)^{\frac{2n+3}{n+1}} \cdot 3d \cdot A^{\frac{1}{n+1}}}{2 \cdot 5(\rho_c^3 - \rho_N^3) \sum\limits_{m=1}^{N} (m)^{\frac{n}{n+1}}} \left(\frac{C^*}{I_n}\right)^{\frac{n}{n+1}} \tag{19}$$

217

It should be noted that this model is restricted to SS creep conditions only, when C^* characterizes the HRR singularity. Most of the tests conducted in this investigation were in the SS regime, hence the model is appropriate.

The following interpretations can be drawn from the model presented in Eq. 22:

1. The crack growth rate is not very sensitive to A for a constant value of n. For this reason, the crack growth rates are expected to be only weakly dependent on temperature, which has been observed for Cr-Mo steels [28].
2. CCGR will be sensitive to the intercavity spacing (2b), the critical cavity size for coalescence ρ_c and the cavity nucleation size ρ_N. If cavities nucleate at GB carbides, then the carbide size can be considered to be ρ_N. therefore, a coarsened GB carbide structure will results in higher crack growth rates.

Although the model does qualitatively predict the trend observed in this study, which is encouraging, it is important to note its limitations. The model assumes an idealized cavitation behavior on a grain boundary along the plane of the crack. It also assumes that cavity growth occurs under steady state stress fields characterized by the HRR singularity, which limits its application to steady-state conditions. Also, the critical cavity radius for coalescence is a constant value, which may be an over-simplification. In reality ρ_c must be stress dependent over a wide range of crack growth rates, and such dependence must be incorporated into the model. Also, under the more realistic transient creep conditions (SSC and TC), the stress fields ahead of the crack tip are relaxing with time and the model must be modified to account for that.

For the Cr-Mo material used in this study, it was observed that the largest cavities ahead of a crack tip were on the order of 70 μm, while the average carbide size ρ_N was on the order of a few microns. If the largest cavities can be taken as ρ_c, then the term $\rho_c^3 - \rho_N^3$ in the model is relatively insensitive to small changes in ρ_N and the crack growth rate will be controlled by ρ_c. The possibility of embrittling species such as tin, antimony, sulphur and phosphorus affecting the critical radius for cavity coalescence needs to be addressed and is being investigated.

Summary and Conclusions

Creep deformation and creep crack growth behavior of ex-service Cr-Mo steel header material were investigated. Two levels of in-service degraded materials were used.

The creep deformation and rupture behavior of the more severely degraded (hot end) material was inferior to the less severely degraded material (cold end) and both ends exhibited inferior properties when compared with virgin material properties. The differences between the two ends were more pronounced at low applied stress levels.

Creep crack growth rates (CCGR) were successfully correlated with the C_t parameter which uniquely characterizes the time rate increase of the crack tip creep zone size. CCGR were higher for the hot end material at low C_t values and essentially the same as the cold end material at higher C_t values. The C_t parameter to some extent normalizes the CCGR behavior with regard to the material state. However, the creep crack growth lives of the two ends are substantially different.

The differences in the creep deformation behavior is attributed to differences in the grain boundary carbide structure. The hot end material had relatively coarser carbides. The small differences in the CCGR behavior between the two ends is most likely a manifestation of the different deformation behavior. The kinetics of grain boundary creep cavitation are not sufficiently enhanced in the hot end material (due to coarser GB carbides) to cause significant differences in the CCGR behavior.

Acknowledgements

The authors wish to acknowledge the support of Electric Power Research Institute through a contract with GA Technologies. Some of the data used in this study was developed at Westinghouse Research Laboratory in Pittsburgh. The assistance of Mr. C.M. Fox in conducting the tests is gratefully acknowledged. Also, helpful discussions with Dr. P.K. Liaw with Westinghouse R&D Center are gratefully acknowledged.

References

1. J.D. Landes and J.A. Begley, in Mechanics of Crack Growth, ASTM STP 590, American Society for Testing and Materials, 1976, pp. 128-148.

2. K.M. Nikbin, G.A. Webster and C.E. Turner, in Cracks and Fracture, ASTM STP 601, American Society for Testing and Materials, 1976, pp. 47-62.

3. A. Saxena, in Fracture Mechanics, Twelfth Conference, ASTM STP 700, American Society for Testing and Materials, 1980, pp. 131-151.

4. D.J. Smith and G.A. Webster, in Elastic-Plastic Fracture, Second Symposium, Vol. I - Inelastic Crack Analysis, ASTM STP 803, American Society for Testing and materials, 1983, pp. I654-I674.

5. H.A. Ernst, in Fracture Mechanics, Fourteenth Symposium - Vol. I: Theory and Analysis, ASTM 791, pp. I499-I599.

6. A. Saxena and J.D. Landes, in Advances in Fracture Research, ICF-6, S.R. Valluri, et.al., eds., Pergamon Press, Vol. 6, 1985, pp. 3977-3988.

7. N.L. Goldmand and J.W. Hutchinson, International Journal of Solids and Structures, Vol. 11, 1975, pp. 575-591.

8. C.F. Shih, Tables of Hutchinson-Rice-Rosengren Singular Field Quantities, Materials Research Laboratory, Brown University, MRL E-147, 1983.

9. J.H. Hutchinson, Journal of Mechanics of Physics and Solids, Vol. 16, 1968, pp. 13-31.

10. J.R. Rice and G.F. Rosengren, Journal of Mechanics of Physics and Solids, Vol. 16, 1968, pp. 1-12.

11. A. Saxena, in Fracture Mechanics, Seventeenth Volume, ASTM STP 905, American Society for Testing and Materials, 1986, pp. 185-201.

12. A. Saxena and P.K. Liaw, "Remaining-Life Estimation of Boiler Pressure Parts - Crack Growth Studies," EPRI CS-4688, Palo-Alto, Ca., July, 1986.

13.   J.L. Bassani, D.E. HAwk and A. Saxena, "Evaluation of the C_t Parameter for Characterizing Creep Crack Growth Rate in the Transient Regime," Third International Symposium on Non-Linear Fracture Mechanics, Knoxville, TN, Oct. 1986.

14.   C. Leung, D.L. McDowell and A. Saxena, "A Numerical Study of Nonsteady State Creep at Stationary Crack Tips," Third International Symposium on Non-Linear Fracture Mechanics, Knoxville, TN, Oct. 1986.

15.   A. Saxena and J. Han, "Evaluation of Crack Tip Parameters for Characterizing Crack Growth Behavior in Creeping Materials," ASTM E24.0807/E24.04.08 Task Group Report, Georgia Institute of Technology, 1986.

16.   "Recommended Practice for Conducting Creep, Creep-Rupture and Stress-Rupture Tests of Metallic Materials," ASTM Standard E139-79, ASTM Book of Standards, Part 10, American Society for Testing and Materials, 1980, pp. 332-347.

17.   H.H. Johnson, Materials Research and Standards, Vol. 5, No. 9, 1965, pp. 442-445.

18.   K.H. Schwalbe and D. Hellman, Journal of Testing and Evaluation, Vol. 9, 1981, pp. 218-221.

19.   A. Saxena, Engineering Fracture Mechanics, Vol. 13, 1980, pp. 741-750.

20.   ASTM DS50, "Evaluation of the Elevated Temperature Tensile and Creep-Rupture Properties of ½Cr-½Mo, 1Cr-½Mo and 1¼Cr-½Mo-Si Steels."

21.   A. Saxena, T.P. Sherlock and R. Viswanathan, in ASME Conference on Remaining Life Prediction of Power Plant Components, Washington, D.C., June 1986, Pergamon Press (in press).

22.   R. Viswanathan, "Microstructural and Miniature Specimen Evaluation for Remaining Life Estimation," 11th International Symposium on Testing and Failure Analysis, Long Beach, CA, Oct. 1985.

23.   M.H. Yoo, C.L. White and H. Trinkaus, in Flow and Fracture at Elevated Temperatures, Proceedings of 1983 ASM Materials Science Seminar, R. Raj, Ed., American Society for Metals, 1985, pp. 349-382.

24.   W.D. Nix and J.C. Gibeling, in Flow and Fracture at Elevated Temperatures, Proceedings of 1983 ASM Materials Science Seminar, R. Raj, Ed., American Society for Metals, 1985, pp. 1-60.

25.   D.S. Wilkinson and V. Vitek, Acta Metallurgica, Vol. 30, 1982, pp. 1723-1732.

26.   J.L. Bassani and V. Vitek, Proceedings of the 9th National Congress of Applied Mechanics - Symposium on Non-Linear Fracture Mechanics, L.B. Freund and C.F. Shih, eds., 1982, pp. 127-133.

27.   J.R. Rice, Acta Metallurgica, Vol. 29, 1981, pp. 675-681.

28.   A. Saxena, J. Han and K. Banerji, "Creep Crack Growth Behavior in Power Plant Boiler and Steam Pipe Steels," EPRI Project RP 2253-10, Georgia Institute of Technology, 1987.

THE EFFECT OF PRESTRAIN ON INTERGRANULAR FRACTURE OF IRON

K. Hashimoto* and M. Meshii

Materials Research Center
and
Department of Materials Science and Engineering
Northwestern University, Evanston, IL 60201

Abstract

The effect of prestrain at ambient temperature on the fracture stress at 77K was investigated in polycrystalline iron. It was found that a very small prestrain increased the fracture stress substantially. The prestraining effect was significant at a stress level below the proportional limit, increased rapidly within the first few tenths of one percent of plastic strain, and reached nearly a saturation level at one percent where the average fracture stress was about 2.3 fold of that of the specimens without prestraining. It is proposed that the heterogeneity in plastic deformation caused by elastic anisotropy between grains is substantial near the macroscopic yield point and produces the compressive residual stress upon unloading from prestraining. Since this residual stress is largest in the region near grain boundaries where the local stress is largest upon reloading, it is responsible for the observed increase in the fracture stress. The present observation indicates that the fracture stress can be substantially changed, if minute straining should occur during setting a specimen into tensile grips.

* Present Address: R & D Laboratories - I, Nippon Steel Corporation, 1618 Ida Nakahara-ku, Kawasaki, Japan

Introduction

The effect of prestrain at ambient temperature on the fracture stress of iron and its alloys at cryogenic temperatures has been investigated previously (1,2,3). Jolley and Hull (4) examined the effect of prestrain on the brittle fracture of Fe-Si alloy and found that fracture stress was increased by prestrain. Furthermore, this increase was equal to the increase in flow stress at room temperature due to the prestrain. On the other hand, Lindley (5) showed opposite prestraining effect in a low carbon steel, where prestrain at room temperature increased the susceptibility to brittleness and 2% prestrained specimen fractured lower stress level than that of a straight test at 77K. These controversial observations of the prestraining effect may be attributed to the difference in the materials. There is, however, little understanding how a small amount of prestrain affects the brittle fracture stress.

It has been demonstrated in earlier investigations (6-8), that a small amount of prestrain at room temperature alters the yielding behavior of iron at lower temperatures. The elastic limit at 77K, for example, decreases with the amount of prestrain at room temperature. This softening effect is usually attributed to the generation of fresh edge dislocations during prestraining. At a low temperature such as 77K, edge dislocations are thought to become mobile at a considerably lower stress than that for screw dislocations.

In the present work, the effect of prestrain is examined when the brittle fracture occurs intergranularly in iron. It is expected that the mechanism involved in intergranular fracture (IGF) is different from those of other fracture modes such as cleavage fracture. The condition of the iron specimens and the test method employed in the present investigation enables us to examine IGF exclusively. The magnitude of the change in fracture stress found in this study is considerably greater than those reported earlier.

In this study, the effects of prestrain at 296K on the intergranular fracture stress of iron were investigated (applying uniaxial tensile stress conditions) at 77K and discussed its phenomenon accounting on the heterogeneity of plastic strain near the grain boundary caused by the prestraining.

Experimental Procedure

The starting material used in the present study was MRC-VP grade iron which contained 40 wt.ppm of sulfur and 350 wt ppm of total substitutional impurity. Iron was swaged to 5mm rod and machined to standard tensile specimen whose gauge length was 12.5mm and diameter was 2.5mm. All specimens were purified in a hydrogen atmosphere which was continuously purified by ZrH_2 at 1133K for 120 hours and then furnace cooled. Earlier works demonstrated that this treatment removes nearly all of carbon and nitrogen (9). The removal of oxygen, however, is not expected to be as complete as carbon and nitrogen (10). The average grain size of the purified specimen was 90 μm and no visible texture was observed. Specimens were chemically polished in a solution of 95%H_2O_2 - 5%HF and rinsed in 50%H_2O - 50%H_2O_2 then washed in methanol before the tensile test.

It has been pointed out previously (11-13), that a small amount of carbon strengthens grain boundary effectively. The complete decarburization of the present specimens, therefore, lowered the IGF stress appreciably so that IGF took place below the yield stress level of single crystals of the same material. The weakening of grain boundaries also occurred by the segregation of sulfur to the grain boundaries. An earlier measurement (14) indicated that the IGF stress was reduced at a rate of 16 MPa/ atomic % sulfur on the grain boundary (monolayer coverage).

Prestrain was introduced by extending a specimen at 296K in an Instron machine with strain rate 1×10^{-4}/sec. Plastic strain was measured as the deviation from the extension of the elastic portion of the stress-strain curve. Then the specimen was subsequently lowered into the liquid nitrogen temperature and was extended again with a strain rate of 2.5×10^{-5}/sec until fracture. All specimens fractured intergranularly without observable plastic strain due to slip at 77K. The yield stress of single crystals grown from the same MRC-VP grade iron was in the range between 500 and 600 MPa at 77K. Tensile stress and tensile strain are used to report the current results throughout this manuscript.

Results and Discussion

Effect of Prestrain

In Fig. 1, typical stress-strain relations are shown to demonstrate the effects of pre-straining at ambient temperature on the fracture behavior at 77K. Specimen A was extended at 77K without prestrain, showing the lowest fracture stress. Specimen B was prestrained 0.12% at 296K and was extended again at 77K. The fracture occurred at a stress level considerably higher than that of specimen A. Specimen C was extended to a plastic strain of 2.17% at 296K and was deformed again to fracture at 77K. The fracture stress increased further but with a considerably smaller rate. It is clear from this illustration that the strengthening effect of prestrain is more than an order of magnitude greater than the work-hardening during prestraining and the effect appears to saturate at a small amount of prestrain. The serrations due to mechanical twinning were rarely observed.

The yield stress of pure iron exhibits strong temperature depen-dependence below ambient tempera-ture (15). The yield stress of the present specimens was around

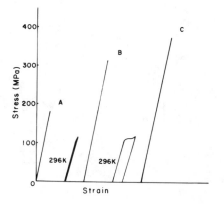

Fig. 1
Typical stress strain curves of polycrystalline iron specimens
A) Tested at 77K without pre-straining
B) Prestrained to 0.12% at 296K and tested at 77K
C) Prestrained to 2.17% at 296K and tested at 77K

100 MPa at 296K. At 77K, the intergranular fracture took place without observable plastic strain in all the specimens tested. The yield stress of single crystals grown from the same MRC-VP grade iron was in the range of 500 ~ 600 MPa. The work hardening rate has considerably smaller temperature dependence than the yield stress as observed in iron single crystals (16).

223

In Fig. 2, the fracture stresses, σ_F is plotted as a function of the amount of prestrain at 296K. It rapidly increases with increasing prestrain for the first few tenths of a percent and reached saturation around 1% prestrain. The increase in the average fracture stress amounts to about 130%. The scatter in the fracture stress data obtained from the prestrained specimens is considerably smaller than those from the specimens without prestrain.

The effect of prestraining at room temperature was significant even when the stress-strain curve did not show any observable plastic strain (below the proportional limit). This effect can be illustrated better in Fig. 3 in which the fracture stresses are plotted against the maximum stress level reached during prestraining. Again the fracture stress is seen to increase with the prestress level. It should be noted that the prestressing to the elastic limit increases the fracture stress by about 150 MPa, doubling the fracture stress. Indeed, Fig. 3 demonstrates that the major part of the strengthening effect takes place below the elastic limit. As the plastic deformation begins to take place heterogeneously in the vicinity of the macro yield stress, the above observation strongly suggests that the local deformation is responsible for the observed strengthening effect.

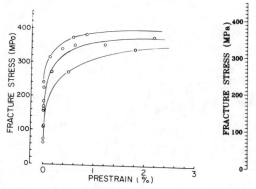 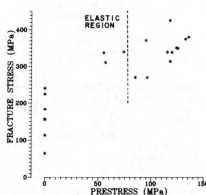

Fig. 2
Variation of fracture stress at 77K with the amount of pre-straining at 296K

Fig. 3
Dependence of fracture stress at 77K on the prestress level at 296K

A Model of Prestraining Effect

It is generally accepted that local dislocation activities initiate brittle fracture. The resulting dislocation configuration, such as dislocation pile-ups against grain boundaries produces stress concentration and microcracks are formed. If a microcrack reaches a critical size which can be evaluated by the Griffith theory (17,18), the crack is no longer stable and propagates to an entire specimen. The strength of a specimen is determined by the weakest element, namely, the highest stress concentration in the specimen. Consequently, the heterogeneity in deformation and microcrack formation is the important feature in determining the fracture stress. Let us examine how a small amount of prestrain at room temperature affects the brittle fracture process.

It has been shown that the elastic interaction stresses which are generated by the neighboring grains having different elastic properties because of orientation difference, play an important role in initiating slip near the grain boundaries (19,20). Furthermore, using the three dimensional Finite Element Method (FEM) analysis, it has been demonstrated that each stress component is not uniformly distributed near the grain boundary. The stress distribution at the vicinity of grain boundary even under uniaxial stress conditions is quite complicated depending on their orientation relationship and constraint conditions of grains (21,22). As these FEM calculations are based on the elasticity theory, it is a general phenomenon in anisotropic elastic bodies including many metals and alloys. The anisotropic factor, $2 \, C_{44}/(C_{11} - C_{12})$, is 2.4 for iron according to the elastic constant data at room temperature (23). Therefore, a significant anelastic effect is expected in iron.

Let us estimate the stresses generated by elastic anisotropy of neighboring grains by using a model of four grains as illustrated in Fig. 4, in which the grains are designated as A,B,C and D. A uniaxial tensile stress is applied to the specimen containing these grains in the direction of the x-axis, and the boundary between A and B grains is perpendicular to this direction.

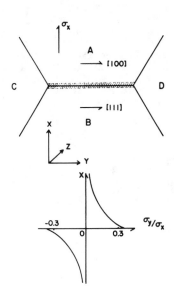

Fig. 4

a) A model of polycrystalline iron; A,B,C and D grains exist under an uniaxial tensile stress condition.
b) The local tensile stress in the y direction – applied stress ratio, σ_y/σ_x is plotted as a function of distance from the grain boundary.

The tensile stress component in the direction of the y axis can be regarded to represent the elastic interaction stresses between A and B grains. Adapting Nyes calculation on the stress distribution near the slip plane (24), the stress component, σ_y at the grain boundary (A/B) will be described as

$$\sigma_y = \pm \, E \, \epsilon_y \, / \, 2 \, (1-\nu^2) \qquad (1)$$

where there is a thin layer of homogeneous grain boundary zone whose Youngs Modulus is E and Poisson's ratio ν . σ_y is the tensile stress component in

225

the y direction at grain boundary. The stress must be calculated in each grain as

$$\epsilon_y^A = S_{21}^A \, \sigma_x, \quad \epsilon_y^B = S_{21}^B \, \sigma_x \qquad (2)$$

where S^A and S^B are the elastic compliance of grain A and B respectively. Let us consider a case in which the y axis coincides with a <100> direction in the A grain and a <111> direction in the B grain as shown in Fig. 4. In the middle region of the grain boundary the influence of the C and D grains can be considered small. Elastic strains at grain boundary ϵ_y^A and ϵ_y^B are calculated applying Hooke's law.

$$\epsilon_y^A = -0.287 \, \sigma_x \, , \quad \epsilon_y^B = -0.0638 \, \sigma_x \qquad (3)$$

where elastic compliances of iron at room temperature are $S_{11} = 0.760 \times 10^{-12}$, $S_{12} = -0.287 \times 10^{-12}$, $S_{44} = 0.892 \times 10^{-12}$ cm^2/dyne (23). The stress component, σ_y is calculated using equations (1) and (3) as

$$\sigma_y = \pm \, 0.714 \, \sigma_x \, / \, 2 \qquad (4)$$

The stress component, σ_y decreases rapidly with distance from the grain boundary as illustrated in Fig. 4b. This estimation indicates that the stress difference between two grains can be reached approximately 0.7 times of applied stress, σ_x. FEM calculations on a Cu-Al bicrystal model (25) have also shown a similar trend of the stress distribution for a grain boundary perpendicular to the direction of applied stress. The preceding calculation demonstrates that large stress heterogeneity due to elastic interaction between grains can exist in polycrystalline iron. These elastic interaction stresses at grain boundary promote local plastic deformation at an applied stress below the macro yield point.

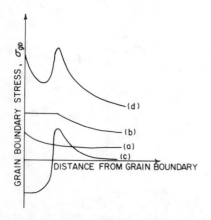

Fig. 5
The grain boundary stress, σ_{gb}, predicted from the proposed model at four stages of prestraining and final straining. The work-hardening effect is neglected in the figure.

a) prestrained slightly at 296K.
b) prestrained beyond the local yield point.
c) after releasing applied stress.
d) strained again at 77K.

The actual stress distribution due to elastic anisotropy of grains can be more complicated and vary from a grain to a grain. The most important case for the intergranular fracture is associated with the boundaries perpendicular to the applied tensile axis. Of all the boundaries, those with the largest interaction stress initiating the intergranular fracture are the current interest. It is also noted in the preceding calculation that the interaction stress is largest at the grain boundary and decays rapidly with the distance from the boundary (Fig. 4b). In the following discussion, the resolved shear stress on the slip system of the largest Schmid factor due to the sum of the interaction stress and the applied stress is referred as the grain boundary stress, σ_{gb} which will cause the fracture at the boundary when it reaches a critical value.

When a specimen is prestrained at ambient temperature, the regions having the highest σ_{gb} will reach the yield point first and plastically deform. Since the yield stress of iron is considerably lower at ambient temperature than at 77K, this condition for the local deformation is reached before σ_{gb} reaches the critical value for the IGF. The local plastic deformation prevents further increase in σ_{gb} (neglecting a small work hardening). Further increase in the applied stress increases the plastic strain and the size of the plastic region (Fig. 5b).

When the applied stress is removed at the end of prestraining, the local plastic strain at the grain boundary region causes the residual stress which is always opposite to the grain boundary stress due to the applied stress (Fig. 5c). When the specimen is deformed again at 77K, the grain boundary stress in this region would be lower than that expected for a specimen without prestraining (Fig. 5d). A higher applied stress must be imposed on the specimen to generate the critical grain boundary stress initiating the intergranular fracture in the prestrained specimen.

The model must be further refined in order to calculate the magnitude of the strengthening effect more accurately. It is, however, useful in understanding the observed strengthening effect semi-quantitatively. The observation that prestraining reduces the scatter in the fracture stress, is also consistent with the proposed model. The strengthening effect takes place in the regions of the highest grain boundary stress. The observed scatter arises from the scatter in grain boundary stress from a specimen to a specimen. Since the prestrain at ambient temperature reduces the grain boundary stress for the subsequent stressing according to its magnitude, it selectively strengthens the weakest spots in a specimen. Similarly, it also predicts that the strengthening effect is considerably larger for the minimum fracture stress (the lower curve in Fig. 2) than the maximum fracture stress (the upper curve).

An important implication of the present observation results from the fact that prestraining within the elastic limit causes a large strengthening effect. Such prestraining can take place unintentionally during specimen handling such as fixing a relatively thin specimen in tensile grips. Indeed, the present investigation was inspired by the discrepancy between the earlier results (14), and the recent results (26). A difference of about 75 MPa in the average fracture stress was observed between two sets of data. The difference was traced to different techniques in griping the specimens. In the former experiment (14), it was carried out at ambient temperature, while it was done at 77K in the latter experiments, resulting in the discrepancy due to an unintentional prestraining effect in the former. The necessity to minimize hydrogen diffusion dictated the specimen gripping in a liquid nitrogen bath and facilitated the detection of the prestraining effect.

Acknowledgement
This work is supported by the MRL program of the National Science Foundation through the Northwestern University Materials Research Center under Contract No. DMR8216972.

References

1. D. V. Wilson and B. Russell; Acta Met. 8, 36 (1960).

2. E. J. Ripling and W. M. Baldwin, Jr.; Trans. Amer. Soc. Metals, 43, 778 (1951).

3. E. T. Wessel; Proc. Amer. Soc. Test. Mat., 56, 540 (1956).

4. W. Jolley and D. Hull; Acta Met., 12, 1337 (1964).

5. T. C. Lindley; Acta Met., 15, 397 (1967).

6. H. D. Solomon and G. J. McMahon, Jr.; Acta Met., 19, 291 (1970).

7. A. Sato and M. Meshii; Phy. Stat. Sol. (9), 28, 561 (1975).

8. R. Kossowsky and N. Brown; Acta Met., 14, 131 (1966).

9. D. F. Stein, J. R. Low, Jr. and A. U. Seybolt; Acta Met 11, 1253 (1963).

10. K. Maruyama, M. Meshii and H. Oikawa; Acta Met 34, 107 (1986).

11. S. Suzuki, M. Obata, K. Abiko and H. Kimura; Trans. ISIJ, 25, 62 (1985).

12. A. Kimura and H. Kimura; Trans. JIM, 25, 411 (1984).

13. K. S. Shin, C. G. Park and M. Meshii; Hydrogen Effects in Metals, TMS-AIME, 209 (1981).

14. K. S. Shin and M. Meshii; Scripta Met. 17, 1121 (1983).

15. J. W. Christian; Met. Trans. 14A, 1237 (1983).

16. D. J. Quesnel, A. Sato and M. Meshii; Mat. Sci. Eng. 18, 199 (1975).

17. A. A. Griffith; Phil. Trans. Roy. Soc.; A221, 163 (1920).

18. K. S. Shin and M. Meshii; Fracture: Interaction of Microstructure, Mechanisms and Mechanics in Fracture, TMS-AIME, 205 (1984).

19. R. E. Hook and J. P. Hirth; Acta Met., 15, 535 (1967).

20. K. Hashimoto and H. Margolin; Acta Met., 31, 773 (1983).

21. H. Miyamoto, M. Shirotori, T. Miyoshi and M. Daito; Journal of the J.S.M.E., 74, 53 (1971).

22. K. Hashimoto and H. Margolin; Acta Met., 31, 787 (1983).

23. J. P. Hirth and J. Lothe; Theory of Dislocations, p. 466 McGraw-Hill, New York (1968).

24. J. F. Nye; Proc. Roy. Soc., A200, 47 (1950).

25. H. Miyamoto, J. Oda and S. Sugimori; The 17th Japan Congress on Material Research-Metallic Materials, 53, (1974).

26. K. Hashimoto and M. Meshii, in this Proceedings.

EFFECTS OF HYDROGEN ON INTERGRANULAR FRACTURE OF IRON

K. Hashimoto* and M. Meshii

Materials Research Center
and
Department of Materials Science and Engineering
Northwestern University, Evanston, IL 60201

Abstract

The effects of hydrogen on the intergranular fracture stress were
studied using polycrystalline iron specimens quenched from high temperatures
in hydrogen atmosphere into 77K and tensile-tested at the same temperature.
The concentration and state of hydrogen were determined utilizing the UHV-
hydrogen analysis system over a temperature range from 77 to 800K. The
results of hydrogen measurements and mechanical tests at 77K were correlated
to obtain the hydrogen concentration dependence of fracture stress. It was
found that solute or weakly trapped hydrogen at dislocation reduced the
fracture stress significantly. A detailed examination of fracture surfaces
and stress-strain curves indicated that the microtwinning took place prior
to fracture. Assuming that hydrogen-induced twinning initiates the
intergranular fracture, the hydrogen concentration dependence of the
intergranular fracture was examined.

*   Present Address: R & D Laboratories - I, Nippon Steel Corporation,
1618 Ida Nakahara-ku, Kawasaki, Japan

Introduction

Hydrogen embrittlement of structural materials such as iron and steel is a serious problem in their application whenever an environmental effect exists. Although many mechanisms involving hydrogen produce an embrittlement effect (1,2), hydrogen-induced intergranular fracture (IGF) is particularly devastating and is the major cause of hydrogen related failure. In this report, the effect of hydrogen on the grain boundary strength will be examined by relating the IGF stress with the concentration of hydrogen.

Recently, the improvements of surface analytical instruments such as Auger electron spectroscopy (AES) and secondary ion mass spectroscopy (SIMS) have made it possible to determine impurity segregation at fracture surface more accurately (3-5). It has been reported that the segregation of sulfur (6-9), phosphorous (7), and oxygen (10-12), at grain boundaries enhances IGF in iron. On the other hand, the presence of carbon (13-15), has been reported to strengthen grain boundary.

Hydrogen trapping by various lattice imperfections such as grain boundaries, dislocations, microvoids and highly strained regions around crack tips have been studied by measurement of diffusion (16-18), internal friction (19,20), hydrogen evolution (21-23), and tritium autoradiography (24), and ion channeling (25). The relation of these hydrogen trapping phenomena to IGF still remains unclear. In the present study, the trapping state of hydrogen in iron was examined by the evolution of hydrogen in an ultra high vacuum hydrogen analysis system, while a specimen was heated up from 77 to 800 K.

It has been known that hydrogen affects the grain boundary strength in two different manners: Hydrogen which exists as solutes or as trapped weakly in the vicinity of grain boundary promotes IGF. On the other hand, hydrogen precipitates at grain boundaries by forming microcracks. The former effect recovers at ambient temperature as hydrogen in this state is highly mobile. The latter effect remains even after hydrogen has diffused out of a specimen, as microcracks do not heal by themselves. While the latter effect has been studied earlier and analyzed by deriving various relations (9,26), the latter effect has not been studied quantitatively, because of a low solubility limit and a high mobility of hydrogen around ambient temperature. Generally speaking, the hydrogen concentration dependence of a mechanical property has been rarely established in the study of hydrogen embrittlement. Such a relation will be established in the present study.

Experimental Procedure

The specimen preparation and tensile test techniques are discussed in the preceding paper (27).

In the present study, hydrogen was charged into iron by means of quenching from a high temperature in hydrogen gas atmosphere. This hydrogen quenching method can introduce a substantial concentration of hydrogen into iron and that the extent of hydrogen trapping is considerably less than that of electro-chemical charging methods which are most commonly applied in hydrogen studies (28). The quenching procedure was as follows: A specimen was kept at 1133K in a dynamic flow of hydrogen gas atmosphere for 1 hour, cooled down to a quenching temperature, T_q where it was kept for 1 hour, and quenched into a $CaCl_2$ - H_2O solution at 243K. After quenching, specimens were chemically polished to remove surface contamination, washed in methanol at 273K, and transferred to a liquid nitrogen bath. In order to minimize the diffusion of hydrogen near room temperature, the time spent for quenching and subsequent handling was kept within 40 seconds. The quenching rate achieved in this method was several thousand degrees K per second.

The concentration of quenched-in hydrogen and the trapping state of hydrogen were examined by determining the thermal evolution of hydrogen from as-quenched specimens in a UHV hydrogen analysis system. The detail description of the UHV hydrogen analysis system was presented elsewhere (28). Hydrogen evolution from the specimens was directly detected by a mass-spectrometer as a function of specimen temperature during isochronal heating from 77 to 800K with a heating rate of 1 K/min. The evolution rate (ppm/min) was calculated from the change in the partial pressure of hydrogen in the system, applying the ideal gas equation, and assuming that the pumping speed for hydrogen by the ion pump was 40 l/sec over the pressure range from 10^{-3} to 10^{-6} Pa. The sensitivity of this system was better than 0.1 ppm/min.

Experimental Results

Fracture Stress

In Fig. 1, fracture stresses, σ_F of both as-quenched, and quenched and aged specimens are plotted against quenching temperature. Each data point represents the average fracture stress of several specimens. The mean deviation is also indicated on the figure. The solid line indicates the average fracture stress value of the specimens quenched and aged at 296K. The broken line in Fig. 1 represents the average fracture stress values of as-quenched specimens. The solid line represents the permanent effect of hydrogen. It has been shown in previous studied (15,26), that this effort does not recover at all near room temperature. The formation of microcracks along grain boundaries has been observed and is suggested to be responsible for this effect. The broken line represents the recoverable effect of hydrogen and recovers quickly at ambient temperature if a small concentration of hydrogen is involved, or is converted to the permanent effect as the fracture stress recovers partially. This effect is thought to be due to highly mobile hydrogen such as solute hydrogen or weakly trapped hydrogen.

Fig. 1

Dependence of the fracture stresses of both as-quenched, and quenched and aged specimens on quenching temperature. Solid line indicates the average fracture stress value of the aged specimens. The broken line represents the average fracture stress values of as-quenched specimens.

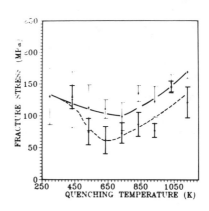

231

Another effect of quenching on IGF stress arises from the fact that the impurity segregation to grain boundaries varies with quenching temperature. Shin and Meshii (9,26), have quantified sulfur and permanent hydrogen effects using Scanning Auger Electron Microscopy and concluded that their effects on fracture stress of iron were found to be independent and additive. Since the amount of sulfur segregation on grain boundary decreases with increasing quenching temperature, its effect on IGF stress is opposite to hydrogen in the quenching temperature dependence. Therefore, in order to obtain the hydrogen effect, the sulfur effect must be subtracted from the observed IGF stress. In Fig. 2 the fracture stress of the reference specimens, which were quenched in argon atmosphere to show only the sulfur effect, is shown along with the fracture stress curve for the as-quenched specimens. The difference between these curves is regarded as the reduction in IGF stress due to quenched-in hydrogen and will be plotted against hydrogen concentration in the subsequent section.

Fig. 2

Dependence of the fracture stress of reference and as-quenched specimens on quenching tempera-ture. The data in reference specimens reported earlier (26), was corrected for the prestrain effect and plotted here.

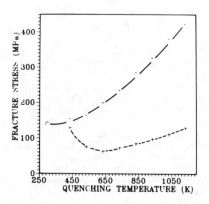

Relationship between Concentration of Quenched-in Hydrogen and Fracture Stress

The concentration of quenched-in hydrogen can be determined from the integration of a hydrogen evolution curve over the entire temperature range. The concentration measured ranged from 55 to 65% of the equilibrium solubility. The values for typical quenching temperature are tabulated in Table I.

Table 1

HYDROGEN CONCENTRATION

Quenching Temperature (K)	Quenched-in Concentration (appm)	Equilibrium Concentration (appm)
733	11	17
933	28	46
1133	50	89

an empirical equation,

$$C_H = 7.62 \times 10^{-4} \exp(-3090/T_q) \qquad (1)$$

was obtained to represent the quenched-in hydrogen concentration as a function of quenching temperature.

The results of the tensile tests and the hydrogen measurement can be combined to examine the concentration dependence of the hydrogen-induced weakening of grain boundaries. In Fig. 3, the decrease in fracture stress is plotted as a function of hydrogen concentration. It can be seen in the figure that a small amount of hydrogen (a few ppm) significantly reduces the fracture stress of iron. Although the rate of hydrogen weakening decreases with increasing hydrogen concentration, the effect does not appear to saturate within the concentration range investigated. The overall concentration dependence can be approximated by a parabolic relation.

Fig. 3

The difference between the fracture stress of as-quenched specimens and reference specimens is plotted as a function of hydrogen concentration.

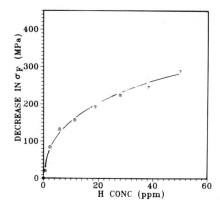

Observation of Stress-Strain Curves and Fractographs

Fig. 4 shows a typical stress-strain curve of an as-quenched polycrystalline specimen. It is noted that the stress-strain curve is serrated at the stress level above 25 MPa. A micrographic examination of IGF surfaces revealed many striations indicating microtwinning. The density of the striations was higher for specimens quenched from a higher temperature. It has also been shown that the presence of hydrogen promotes twinning at 77K in iron single crystals quenched from high temperatures in hydrogen atmosphere (29).

Fig. 4

Stress-strain curve at 77K for a polycrystalline specimen quenched from 733K in hydrogen atmosphere.

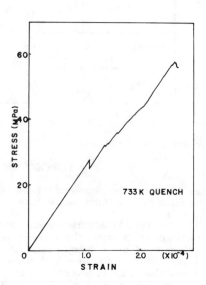

It is clear that hydrogen interacts with dislocations and promotes twinning in iron at 77K. When the propagation of twinning is intercepted by a grain boundary, a micro crack nucleates at the boundary. If a micro-crack is larger than a critical size for a given condition, it propagates along grain boundaries, causing IGF in iron.

Discussion

Twin Induced Intergranular Fracture

Several mechanisms have been proposed to understand why hydrogen reduces the grain boundary strength. i) Cluster model (31); mobile hydrogen diffuses to grain boundaries and precipitates to form microvoids on them. Those microvoids reduce the fracture stress level depending on their sizes. ii) Lattice decohesion model (32,33); atomic hydrogen segregating to grain boundaries weakens the iron-iron bonding at the grain boundaries, facilitating intergranular fracture at a lower stress level. iii) Slip induced IGF (34); although iron fracture in a brittle manner at low temperature, a small amount of plastic deformation is known to take place in the vicinity of fracture surface. Recently Raman and Kamar (10) observed that intergranular cracks are nucleated by the interaction of slip bands initiating from the surface with the oxygen segregated grain boundaries. iv) Twin induced IGF; McMahon (13) concluded that most of the intergranular cracks were nucleated by twins striking grain boundaries at 77K in iron which contained a small amount of carbon and oxygen. The cluster model is thought to cause the permanent hydrogen effect since many intergranular cracks were observed after sufficient aging at room temperature. However, in the case of the recoverable effect, hydrogen is thought to exist near the grain boundary as solutes or weakly trapped hydrogen. The decohesion model can account for the recoverable hydrogen effect. It has been suggested in an earlier report (30), that hydrogen is not trapped at grain boundaries,

instead, hydrogen is still in the state of atomic hydrogen as it is
quenched-in. The serrations in the stress-strain curves and micrographs of
fracture surface suggest that the twinning takes place before IGF and is
related intimately to the initiation of IGF.

In order to examine the effect of hydrogen on twinning, two factors
have to be considered; those are the stress to initiate twinning and the
amount of twinning. In Fig. 5, the twinning initiation stress is plotted
against quenching temperature for both single and polycrystalline iron
specimens. The twinning initiation stress was defined by the first
serration observed on the stress-strain curve. In the case of single
crystal, it is clear that increasing quenching temperature, namely
increasing hydrogen concentration, the twinning initiation stress decreases
from 500 MPa for 833K to 130 MPa for 1133K. These single crystal data
suggest that dissolved hydrogen interacts with dislocations and promotes
twinning at 77K. Furthermore, the initiation stress depends strongly on the
concentrations of hydrogen. In the case of polycrystalline specimens, this
trend is not clear in Fig. 5, the twinning initiation stress of
polycrystalline specimens does not vary significantly with quenching
temperature. At lower quenching temperatures, there is the tendency that
the twinning initiation stress, increases with decreasing quenching
temperatures. The weak temperature dependence of the twin initiation stress
in polycrystalline specimens may be attributed to the saturation effect of
hydrogen in the vicinity of the grain boundaries where the twinning occurs.
The stress gradient discussed in the preceding paper (27), may also be an
important factor to be considered. The grain boundary stress should lower
the twinning initiation stress in polycrystalline specimens; however, the
large difference between single crystals and polycrystalline specimens
suggests a higher concentration of hydrogen near the grain boundaries.
Polycrystalline specimens which quenched from a high temperature in hydrogen
atmosphere show many twinning prior to IGF.

In Fig. 6 the average twin strain is plotted as a function of quenching
temperature. Although a wide scatter is noted, the average twin strain
increases with quenching temperature, indicating the volume of the twinned
increases with hydrogen concentration.

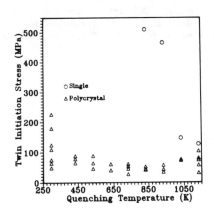

Fig. 5

Twinning initiation stress plotted
as a function of quenching
temperature T_q.

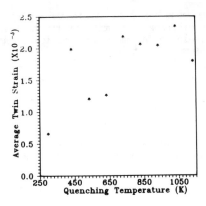

Fig. 6

Average twin strain plotted as
a function of quenching
temperature T_q.

Fig. 7

The relationship between the normalized twin strain and logarithm of hydrogen concentration.

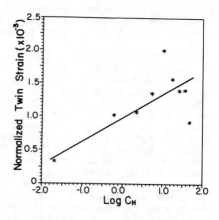

Since the twinning strain is influenced by the fracture stress of a specimen, it should be normalized with a common stress for a better comparison. In Fig. 7, the twinning strain at 100 MPa is plotted as a function of hydrogen concentration estimated from quenching temperature using Eg. (1). As this method of evaluation of twinning strain removes the effect of scattering in fracture stress, the hydrogen concentration dependence of twinning trend is more clearly seen in Fig. 7 where the normalized twinning strain is plotted against the logarithm of the hydrogen concentration. It should be noted that the twinning strain is proportional to the volume fraction of the twinned crystal.

Effect of Hydrogen on Intergranular Fracture Stress

An empirical relation can be obtained from Fig. 7 between the volume fraction of the twinned crystal and hydrogen concentration as

$$V_T = k_1 \log C_H. \qquad (2)$$

Numerous twinning events were observed before IGF finally occurred. The observation suggests that the collision of microtwins at grain boundary is likely to determine the critical crack size, c for IGF. Therefore, let us assume a relation,

$$c = k_2 V_T^{2} \qquad (3)$$

where k_2 is a constant.

Combining equations (2) and (3), the crack size c cab be written as a function of hydrogen concentration C_H.

$$c = k_1^2 k_2 (\log C_H)^2 \qquad (4)$$

According to the classical Griffith theory, the brittle fracture stress, σ is a function of crack size, c as

$$\sigma = (2ES/\pi c)^{1/2} \qquad (5)$$

236

where E is Youngs modulus and S is surface energy. Substituting Eq. (4) into Eq. (5), the fracture stress σ_F can be expressed as a function of the concentration of hydrogen C_H.

$$\sigma = (2ES/\pi)^{1/2} (k_1^2 k_2)^{1/2} (\log C_H)^{-1} \qquad (6)$$

In the earlier section, the decrease in fracture stress, $\Delta\sigma_F$ is defined by the difference between reference and hydrogen quenched specimens. For a reference specimen of a given T_q, fracture stress σ_S will be

$$\sigma_S = (2ES/\pi)^{1/2} (k_1^2 k_2)^{1/2} (\log C_S)^{-1} \qquad (7)$$

where σ_S is fracture stress of the reference specimen and C_S is the hydrogen concentration at 296K. Therefore, the decrease in fracture stress $\Delta\sigma_F$ can be described as

$$\Delta\sigma_F = \sigma_S - \sigma_F = (2ES/\pi)^{1/2} (k_1^2 k_2)^{-1/2} \left\{ (\log C_S)^{-1} - (\log C_H)^{-1} \right\} \qquad (8)$$

In Fig. 8, a linear relationship is found to support this equation. However, the line fails to go through the origin. This may be caused by an underestimation of C_S. The extrapolation of Eq. (1) to the room temperature may not be accurate, since the presence of hydrogen traps can significantly increase the concentration of hydrogen at room temperature.

Fig. 8

The relationship between the decrease in fracture stress $\Delta\sigma_F$ and hydrogen concentration $\{(\log C_S)^{-1} - (\log C_H)^{-1}\}$.

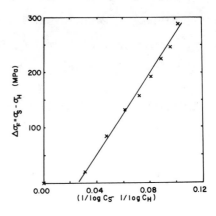

237

Acknowledgement

The authors wish to thank Messrs. Y. Z. Shen, J. Koike and Mr. J. Watson and Dr. K. Ono for their contributions in various phases of the present investigation. The research was supported by the MRL program of the National Science Foundation through the Northwestern University Materials Research Center under Contract No. DMR 82-16972.

References

1. Hydrogen Degradation of Ferrous Alloys Eds. R. A. Oriani, J. P. Hirth and M. Smialowski, Noyes Publication, Park Ridge, NJ (1985).

2. J. P. Hirth; Met. Trans., 11, 861 (1980).

3. D. F. Stein, A. Joshi and R. P. Laforce; Trans. ASM., 62, 776 (1969).

4. K. Yoshino and C. J. McMahon, Jr.; Met. Trans., 5, 363 (1974).

5. H. Fukushima and H. K. Birnbaum; Acta Met., 32, 851 (1984).

6. C. Pichard, J. Rieu and C. Goux; Met. Trans., 7A, 1811 (1976).

7. P. V. Ramasubramanian and D. F. Stein; Met. Trans., 4, 1735 (1973).

8. S. M. Bruemmer, R. H. Jones, M. T. Thomas and D. R. Baer; Scripta Met., 14, 137 (1980).

9. K. S. Shin and M. Meshii; Fracture: Interactions of Microstructures, Eds. J. M. Wells and J. D. Landes, Mechanisms and Mechanics, 205 (1985).

10. A. Kumar and V. Raman; Acta Met., 29, 1131 (1981).

11. H. Matsui and H. Kimura; J. Japan Inst. Metals, 47, 294 (1983).

12. K. Maruyama, M. Meshii and H. Oikawa; Acta Met., 34, 107 (1986).

13. C. J. McMahon, Jr.; Acta Met. 14, 839 (1966).

14. A. Kimura and H. Kumura; J. Japan Inst. Metals, 47, 807 (1983).

15. K. S. Shin and M. Meshii; Fracture:interaction of microstructures, mechanism and mechanics 195, (1985).

16. O. D. Gonzales; Trans. AIME, 245, 607 (1969).

17. K. Kiuchi and R. B. McLellan; Acta Met., 31, 961 (1983).

18. A. J. Kamnick and H. H. Johnson; Met. Trans., 5, 1199 (1974).

19. R. Gibala; Trans. AIME, 239, 1574 (1967).

20. K. Takita and K. Sakamoto; Scripta Met. 10, 399 (1976).

21. G. M. Evans and E. C. Rolason; J. of Iron and Steel Inst., 207, 1484 (1969).

238

22. T. Takeda; Proc. JIMIS-2, Hydrogen in Metals, Suppl. to Trans. Japan Inst. Metals., 21, 237 (1979).

23. W. Y. Choo and J. Y. Lee; Met. Trans. A, 13A, 135 (1983).

24. T. Asakoka, C. Dagbert, M. Aucouturier and J. Gallard; Scripta Met. 11, 467 (1977).

25. S. M. Myers, S. T. Picraux and R. E. Stoltz; J. Appl. Phys., 50, 5710 (1979).

26. K. S. Shin and M. Meshii; Acta Met., 31, 1559 (1983).

27. K. Hashimoto and M. Meshii, published in this Proceedings.

28. K. Hashimoto and M. Meshii; Scripta Met., 19, 1075 (1985).

29. K. Hashimoto and M. Meshii; Strength of Metals and Alloys Eds. H. J. McQueen et. Pergamon, 379, (1985).

30. K. Hashimoto and M. Meshii; Proc. JIMIS-4 Grain Boundary Structure and Related Phenomena (1986), Suppl. Trans. Japan Inst. Metals p. 781.

31. C. A. Zapffe and C. E. Sims; Trans AIME, 145, 225 (1941).

32. A. R. Toriano; Trans. ASM 52, 54 (1960).

33. R. Gibala; Trans. AIME 239, 1574 (1967).

34. R. A. Oriani and Josephic; Acta Met. 22, 1065 (1974).

EFFECT OF CYCLIC LOADING ON FRACTURE

TOUGHNESS OF A MODIFIED 4340 STEEL

*J. D. Landes and **P. K. Liaw

*American Welding Institute, Route 4, Box 90, Louisville, TN 37777
**Westinghouse R&D Center, 1310 Beulah Road, Pittsburgh, PA 15235

Abstract

Structural components often encounter complex load histories during service; cyclic loading can be encountered during the actual fracture process. Previous results have shown that cyclic loading can alter the fracture toughness of steels from the values measured during a standard monotonically loaded test. In this investigation, load cycles were applied to a modified 4340 steel during the development of ductile fracture. The cyclic loading was imposed under controlled displacement conditions with two basic formats, one simulating more a load-controlled situation (labeled ratcheting crack) and another simulating elastic displacement control (labeled elastic dominance). Results were analyzed using a modified J-R curve approach for ductile fracture combined with a da/dN versus ΔJ for fatigue crack growth. The crack extension measured was greater than that which would result separately from either ductile fracture or fatigue. Therefore, the observed crack extension was thought to have both a monotonic and a cyclic component. A model which proposed to predict crack extension from a linear combination of both components did not adequately account for all of the observed growth. A second model containing crack growth due to an interaction of both was proposed. The fracture morphology was investigated to explore any correlation between fracture surface and crack growth response.

Introduction

The life expectancy of a structure subjected to cyclic loading can be evaluated using the fracture mechanics approach. This evaluation is often made by applying fatigue crack growth calculation until the point of fracture toughness is reached, Figure 1.[1,2] The toughness result used in such an analysis, however, is probably generated by a standard test which uses monotonic loading rather than cyclic to measure the toughness value, Figure 1. Previous work has shown that a defect growing to the toughness level during cyclic loading may not fail at that level as predicted. Rather, the toughness may be changed by the cyclic loading imposed during the toughness test. Sometimes the apparent cyclic toughness is increased over that of the monotonic toughness, and sometimes it is decreased, Figure 2.[3,4] Structural components may receive a complex cyclic load history during service; therefore, a straightforward application of fatigue crack growth up to the fracture toughness point may not correctly predict failure.

The examples cited above are taken from the linear elastic fracture mechanics regime. When the process of loading to failure involves elastic-plastic deformation, the fracture is often ductile and involves a process rather than an event.[5] The process of fracture may then be influenced by the cyclic loading. The ductile fracture process is often characterized by the crack growth resistance curve, R curve, Figure 3. Previous work had looked at the influence of cyclic loading on the R-curve development and concluded that cyclic loading increased the rate of crack extension. It was suggested that the effect could be represented by a linear combination of a monotonic contribution to the R curve and a cyclic contribution; however, no conclusive quantitative results were presented which could either verify or disapprove this suggestion.[1,6]

The objective of this paper is to further examine the effect of cyclic loading on the ductile fracture development. Specifically, data were gathered to allow study of quantitative models such as the suggested one involving a linear combination of cyclic and monotonic effects. A modified 4340 steel was used; this steel had been used previously to study the effects of load history on fracture for a number of postulated service conditions.[6,7,8] In this study, ductile fracture behavior was developed with a loading pattern that had superimposed cyclic load excursions. Results were examined from a fatigue crack growth point of view as well as a fracture toughness point of view.

Background

Previous studies of the effects of high load cycles during the ductile fracture process had taken the point of view that a basic R-curve fracture process was being developed and the cyclic loading altered it in some manner. An alternate point of view could be taken that the cyclic loading was basically a fatigue crack growth process which was being altered by progressive opening of the crack (a ratcheting process).

The previous work, which looked at the process from the toughness point of view, used a displacement control where the specimen was loaded, unloaded and reloaded to prescribed values of displacement.[1] The R-curve component of crack growth was analyzed by taking an envelope monotonic loading curve, which included the maximum displacement point of each cycle, and analyzing it using a J-R curve procedure. The result of a monotonic loading case with no cyclic component (virgin specimen) was compared with one that had cyclic loading. The observed effect was that crack extension

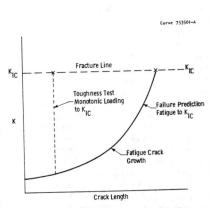

Curve 753501-A

K_{IC} ── × ──── Fracture Line ──── × ── K_{IC}

Toughness Test
Monotonic Loading
to K_{IC}

Failure Prediction
Fatigue to K_{IC}

K

Fatigue Crack
Growth

Crack Length

Fig. 1—Schematic contrasting toughness testing with failure prediction

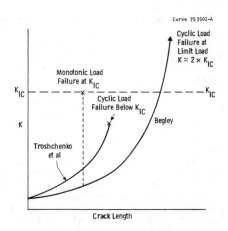

Curve 753502-A

Cyclic Load
Failure at
Limit Load
$K \approx 2 \times K_{IC}$

Monotonic Load
Failure at K_{IC}

K_{IC} ──── × ──── K_{IC}

Cyclic Load
Failure Below K_{IC}

K

Begley

Troshchenko
et al

Crack Length

Fig. 2—Schematic showing results of fatigue to failure tests

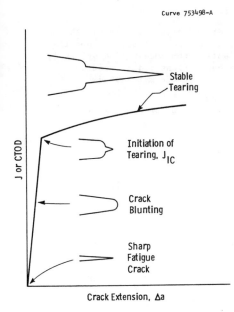

Curve 753498-A

J or CTOD

Stable
Tearing

Initiation of
Tearing, J_{IC}

Crack
Blunting

Sharp
Fatigue
Crack

Crack Extension, Δa

Fig. 3—Ductile fracture development on an R curve

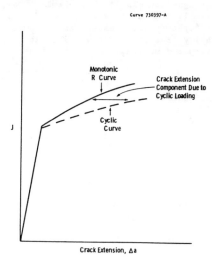

Curve 730397-A

Monotonic
R Curve

Crack Extension
Component Due to
Cyclic Loading

J

Cyclic
Curve

Crack Extension, Δa

Fig. 4—Schematic showing linear sumation principle for developing a cyclic R curve

243

was increased significantly for the cyclicly loaded case as compared with the monotonic case, Figure 4. The suggestion was made that the monotonic R curve would be used to predict the result from the cyclic loads by linearly adding a cyclic component. The cyclic component is determined from a da/dN versus ΔJ fatigue crack growth rate analysis.

Starting from the fatigue crack growth point of view, where crack growth rate, da/dN, is plotted as as a function of ΔK, it has been long observed that the normal power law region of the curve[9] may be altered at high ΔK.[10] A region of accelerating crack growth rate is often observed and attributed to an effect brought on by the onset of fracture. For materials which undergo large plastic effects at high ΔK, Dowling and Begley[11] suggested that the effect may be more a failure of the K parameter to adequately characterize the crack tip field zone. Therefore, with a proper characterizing parameter like ΔJ, the effect of accelerating crack growth may not be observed. Dowling[12] showed that the power law region could be extended to very high rates of crack growth by using ΔJ rather than ΔK, Figure 5. The cyclic loading that Dowling used, however, was carefully controlled so that there was no ratcheting of the end point displacements, essentially ensuring that there was no monotonic load component.[11] This careful control would not necessarily exist for a component under service loading which may receive contributions of both monotonic and cyclic loading during the fracture process. When viewed from a fatigue crack growth point of view, this would result in the accelerating growth rate curve that is observed at the high growth rate end. The amount of acceleration, however, would not be a reproducible material property that can be measured from an ordinary fatigue crack growth rate test. Rather, it would depend on the size of the monotonic component of crack extension and could only be measured from a cyclicly loaded fracture test such as the type used in this study.

Experimental Procedure

The tests were conducted with a modified 4340 steel (given the designation 4335). The chemical composition and mechanical properties are listed in Table 1. It differs slightly from modern 4340 by having a higher nickel content and a lower carbon content and is used in heavy section forgings. The microstructure of the material is tempered martensite.

TABLE 1 - PROPERTIES OF THE MODIFIED 4340 STEEL

A. Chemistry:

C	Mn	P	S	Si	Cr	Ni	Mo	V
0.32	0.69	0.007	0.005	0.26	0.91	2.72	0 42	0.09

B. Room Temperature Tensile Properties:

Yield Strength MPa	(ksi)	Tensile Strength MPa	(ksi)	Percent Elongation	Percent RA
1041	(151)	1121	(163)	16	53

Curve 753506-A

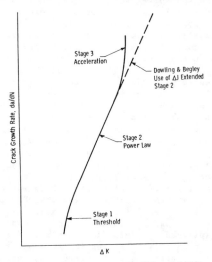

Crack Growth Rate, da/dN

Stage 3
Acceleration

Dowling & Begley
Use of ΔJ Extended
Stage 2

Stage 2
Power Law

Stage 1
Threshold

Δ K

Fig. 5 — Schematic showing extrapolation of fatigue crack growth
rate curves to high Δ K

Curve 753505-B

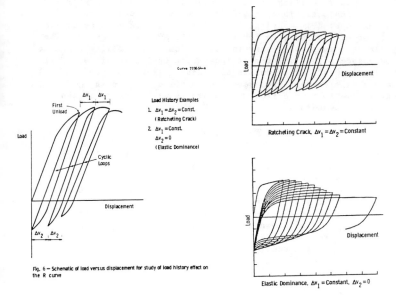

Load

Displacement

Ratcheting Crack, $\Delta v_1 = \Delta v_2 =$ Constant

Load

Displacement

Elastic Dominance, $\Delta v_1 =$ Constant, $\Delta v_2 = 0$

Fig. 7 — Examples of Ratcheting Crack (RC) and Elastic
Dominance (ED) load histories used for developing
cyclic R curves

Curve 729694-A

Δv_1 Δv_1

First
Unload

Load History Examples

1. $\Delta v_1 = \Delta v_2 =$ Const.
   (Ratcheting Crack)

2. $\Delta v_1 =$ Const.
   $\Delta v_2 = 0$
   (Elastic Dominance)

Load

Cyclic
Loops

Displacement

Δv_2 Δv_2

Fig. 6 — Schematic of load versus displacement for study of load history effect on
the R curve

245

The specimen geometry used was the compact type with width 51 mm (2 inch) and thickness 25.4 mm (1 inch). This geometry is sometimes designated 1T-CT; the specimens were not side grooved.

The objective of the tests was to develop ductile fracture while at the same time imposing large cyclic loops in order to simulate something of service loading histories. Under service loading conditions, the plasticity should be confined to a local plastic zone region which is surrounded by elastic material. This would create boundary conditions on the loading which are more displacement controlled than load controlled. Therefore, in these tests, the specimens were loaded to prescribed displacement values rather than to prescribed loads. Typically, the tests were loaded to a given maximum displacement, unloaded, to a minimum displacement and then reloaded to a second maximum. The progression is shown in Figure 6. The loading for all of the specimens was done in a systematic way where the increments of maximum displacement and of minimum displacement were constant for each cycle. Two load patterns were used in this series of tests. In the first, the maximum and minimum increments were equal so that the cycle loops were relatively constant but systematically increased in the mean level of displacement. This was called the 'ratcheting crack' in that the crack was progressively opened. A second type of loading, called 'elastic dominance,' had a progressively increasing maximum displacement increment but was always unloaded to zero displacement. This simulated the case where the elastic boundary had such a large effect that the material is always returned to its original starting strain level upon unloading. The other variable was the size of the displacement increments, hence, the total number of cycles imposed on the specimen during the process of fracture.

Examples of the actual load displacement records are shown in Figure 7 for both ratcheting crack and elastic dominance. A matrix showing all of the conditions tested is given in Table 2. Two specimens were tested in a 'virgin' condition, that is with no cyclic loading, to provide a baseline for the ductile R-curve behavior. The other specimens were tested under the condition of ratcheting crack or elastic dominance with displacement increments that were either 0.127 mm (0.005 in) or 0.025 mm (0.001 inch). This was five to one difference in displacement increment and resulted also in an approximately one to five difference in cycles needed to reach the same total displacements at the end of the test.

Analysis Technique

The analysis of the data was based upon an R-curve evaluation where a fracture mechanics parameter is plotted versus crack extension. The parameter used here was the modified J parameter of Ernst, J_m.[13] The crack growth was looked upon as having a monotonic loading component and a cyclic component. A fracture mechanics parameter was needed for each. For the R-curve analysis, a load displacement envelope was drawn and the modified J determined from area under this curve, Figure 8.[13] For the cyclic crack growth component, area was taken under each individual cyclic loop so that a ΔJ could be determined,[12] Figure 9.

246

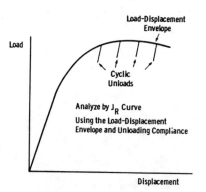

Curve 729652-A

Fig. 8 — Load history effect-analysis by R curve

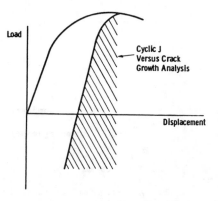

Curve 729653-A

Fig. 9 — Load history effect - analysis by cyclic J

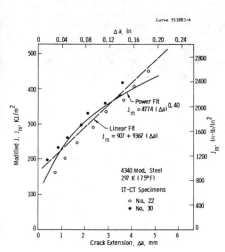

Curve 753883-A

Fig. 10 — Modified J versus crack extension for virgin specimens

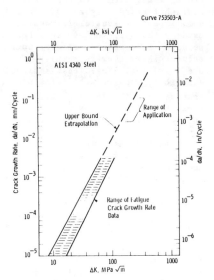

Curve 753503-A

Fig. 11 — Fatigue crack growth rate data showing extrapolation for R curve model

247

TABLE 2 - TEST MATRIX

Spec No.	Test Type	v_o	Δv	Total Cycles
22	virgin	–	–	–
30	virgin	–	–	–
23	ratcheting	0.89 mm	.127 mm	10
24	crack	1.14	.127	7
25	'	1.14	.127	8
28	'	1.27	.127	7
26	'	0.89	.025	46
31	'	1.27	.025	31
27	elastic dominance	0.89	.127	10
32	elastic dominance	0.89	.025	46

1T-CT Compact Specimen

W = 50 mm
B = 25 mm

The evaluation of monotonic fracture in the R-curve format is done by associating a crack extension value with each value of J_m. J_m is determined from area under the load versus displacement curve and formulas given in reference 13. Crack extension is determined from the elastic unloading slope as in the J-R curve test method.[14]

For the two 'virgin' specimens which had no component of cyclic growth associated with them, the presentation of results was straightforward, Figure 10. As can be seen in this figure, there is some amount of scatter in results. Since these results were to be used as the basic monotonic crack growth component in the model to predict the combined effects, some representation had to be made to reflect their average result. To do this two curve fits were made, one a straight line, least squares curve fit in the manner of the classic J_{IC} analysis.[15] The second curve fit used was the power law proposed for the new J_{IC} analysis.[16] Visually, the straight line fit seems to better represent the full range of R curve so this was used in the model to predict results.

The fatigue component analysis was based upon standard da/dN versus ΔK results. In this case, a ΔJ was determined from the area under a single loop, Figure 9. The ΔJ can be converted to an equivalent ΔK for associating it with a fatigue crack growth rate value. Two problems had to be solved in developing these data. First, there were no da/dN versus ΔK results for this material; data from another heat of 4340 steel were used. Secondly, the ΔK values associated with the calculated ΔJ were all large, falling above the range of any measured results. To apply these data, a straight line extension of the upper bound power law region was made, Figure 11. This estimation of fatigue crack growth rate behavior, although perhaps leading to a slight overestimate of the fatigue contribution, was consistent with general results presented by Rolfe and Barsom for martensitic steels.[10]

From the virgin J-R curve and the fatigue crack growth rate, a combination could be made in some manner to represent the total contributions of the monotonic part and the cyclic part. As suggested previously, an obvious first attempt would be to add them linearly. The prediction from the linear model could then be compared to the measured results.

Results

Only four sets of results were chosen for presentation to simplify the discussion. Although more tests were conducted, the cases chosen cover all of the variables involved and typically represent the results of all the tests. The J-R curve results are plotted in Figure 12. These cover the ratcheting crack and elastic dominance cases; for each case, one example of the larger displacement increment (fewer cycles) and one of the smaller displacement increment (greater cycles) are shown. Also included for comparison is the straight line fit of data in Figure 10 labeled 'best fit virgin.' As can be seen in Figure 12, the results fall on or near the fit line and follow its trend initially. As the cyclic loading components accumulate, the results deviate from the fit line by exhibiting additional cracking. This trend was observed for every specimen tested.

The results can be plotted in a da/dN versus ΔK format, Figure 13; this shows that crack growth per cycle is greater than the amount that would be predicted by a straight line extension. Therefore, the assumption presented that the crack extension contains combined components from the monotonic and cyclic loading appears reasonable. The success of predicting the results by combining both components in some fashion is examined in the next section.

Predictive Models

The first attempt to develop a model is through a linear combination of both monotonic and cyclic components. The best fit virgin line was used to predict the monotonic component. To account for some initial scatter, the fit curve was shifted, while keeping constant slope, to pass through the first point. This point represents the initial loading before cycling and therefore is representative of the monotonic R curve of the material. The cyclic component is determined by taking the fatigue crack growth rate from the upper bound extension of the da/dN versus ΔK 4340 steel results. The calculated ΔJ is converted to an equivalent ΔK ($\Delta K = \sqrt{E\Delta J}$) and the corresponding da/dN determines the amount of crack growth on that cycle. For each succeeding cycle, the fatigue component is determined in the same way and accumulated ($\Delta a_f = \sum_N [da/dN]$). The linear method of prediction then adds the Δa_f to the Δa on the best fit line at the appropriate J_m.

The results of the linear prediction are given in Figure 14 for the ratcheting crack and in Figure 15 for the elastic dominance. In every case shown, the linear prediction follows results fairly well for the first few points but fails to follow at larger Δa. It is obvious from these that a linear combination of results is not sufficient. A synergism may exist between the two components of crack extension which results in greater crack extension than predicted by a linear combination of the two. In order to make a reasonable prediction of this effect, some mechanistic model would be helpful. Lacking that, a simple model which proposes a third interactive

249

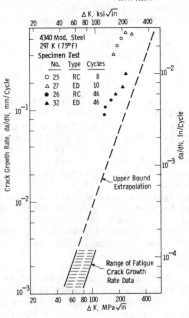

Curve 753507-A

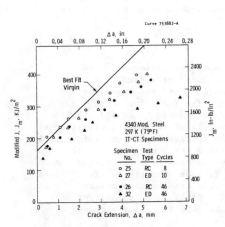

Curve 753882-A

Fig. 12—Modified J versus crack extension for RC and ED specimens

Fig. 13—R curve data interpreted as fatigue crack growth

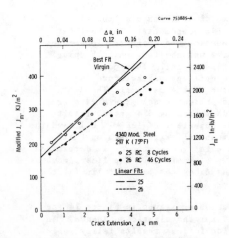

Curve 753885-A

Fig. 14—Modified J versus crack extension showing linear predictions for RC specimens

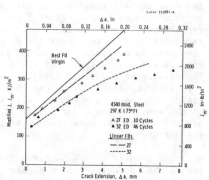

Curve 753881-A

Fig. 15—Modified J versus crack extension showing linear predictions for ED specimens

component of crack growth, one which fits the last point for each
specimen, was used, Figure 16. The results presented in Figures 14
and 15 were replotted in Figures 17 and 18 using this model. The
results, although looking much better, are not completely satisfactory
in that no rationale for choosing such an interactive term can be
given.

Fractographic Results

Fracture morphology was carefully examined using scanning
electron microscopy (SEM). Following fracture toughness tests,
fracture surfaces were heat tinted to determine final crack lengths.
Heavy oxidation was thus observed on fracture surfaces prior to SEM
examinations; fracture surfaces were electrolytically cleaned in an
ENDOX solution.[17]

The SEM photos of virgin, RC and ED specimens are presented in
Figures 19-21, respectively. In each figure, low, intermediate, and
high magnification photos were included. In the virgin material,
dimpled fracture was found to be the dominant mode of fracture, Figure
19c. Similarly, dimpled fracture was also observed to be
characteristic of RC and ED specimens (Figures 20C and 21C). The mode
of fracture in each specimen was found to be insensitive to crack
length. However, the size and the density of dimples in the specimens
with greater number of loading cycles (specimens 26 and 32) seemed to
be smaller than those with fewer number of loading cycles (specimens
25 and 27).

At low and intermediate magnifications, striated ridges were
readily visible in RC and ED specimen, while no clear ridges were
present in the virgin material.

The quantitative measurements of the average spacings of ridges
are listed in Table 3 for RC and ED specimens. Note that the spacing
of ridges in each specimen is insensitive to crack length. In both
specimens, the spacing of ridges seems to be comparable to the average
crack growth rate. These results suggest that these ridges in the RC
and ED specimens may be associated with the cyclic loadings during
fracture toughness testing. Furthermore, each ridge seems to
correpond to each loading cycle.

TABLE 3 - CRACK GROWH RATES AND SPACING OF RIDGES

Specimen Number	Test Type	Average Crack Growth Rate (da/dN) (μm/cycle)	Spacing of Ridges (μm)
25	RC	550	470
27	ED	500	436
26	RC	110	175
32	ED	200	225

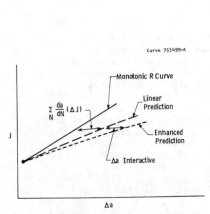

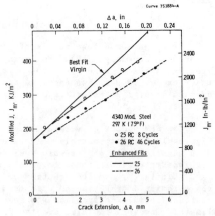

Fig. 16—Schematic of linear and enhanced fit cyclic R curve predictive schemes

Fig. 17—Modified J versus crack extension showing enhanced fit predictions for RC specimens

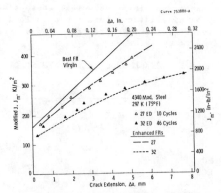

Fig. 18—Modified J versus crack extension showing enhanced fit predictions for ED specimens

252

Stable Crack Growth Region	Stable Crack Growth Region
(a) Low Magnification	(b) Intermediate Magnification

Crack Growth
Direction

(c) High Magnification

Fig. 19 — Fractography of virgin material

(a) Low Magnification

(b) Intermediate Magnification

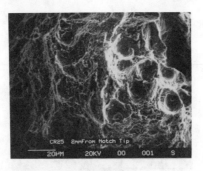

Crack Growth
Direction

(c) High Magnification

Fig. 20 — Fractography of ratcheting crack (RC) test type (specimen CR25)

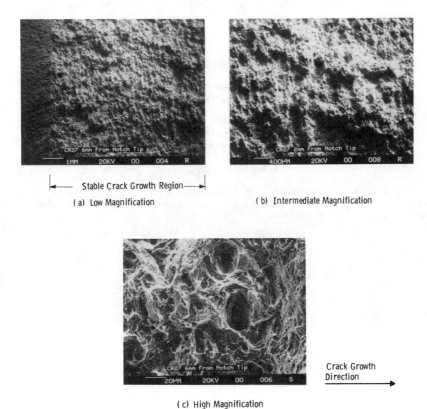

Stable Crack Growth Region

(a) Low Magnification

(b) Intermediate Magnification

Crack Growth
Direction

(c) High Magnification

Fig. 21 — Fractography of elastic dominance (ED) test type (specimen 27)

Conclusions

1. Failure during cyclic loading may not always be adequately predicted by assuming that a crack grows under fatigue until it reaches a fracture toughness level measured by a monotonic load test. The cyclic loading has been found to alter the toughness behavior of some steels.

2. For ductile fracture, R curves developed during cyclic loading appear to combine a monotonic and cyclic component of crack growth.

3. A model using a linear combination of monotonic and cyclic crack growth fell short of accurately predicting the observed crack growth.

4. The addition of a third, interactive component of crack growth, to the model predicts results much better, although no mechanistic rationale for this third term could be given.

5. Dimpled fracture was found to be characteristic of tested specimens regardless of loading history.

6. Striated ridges were observed on the fracture surfaces of the cyclicly loaded specimens; the spacing of the ridges was of the order of the crack growth per loading cycle.

Ackowledgement

M. G. Peck of Westinghouse R&D Center assisted with the experimental aspects of this work. A. Karanovich and P. Yuzawich assisted with the study of the fracture surfaces.

REFERENCES

1. J. D. Landes and D. E. McCabe, 'Load History Effects on the J-R Curve,' Elastic-Plastic Fracture: Second Symposium, Volume II - Fracture Resistance Curves and Engineering Applications, ASTM STP 803, C. F. Shih and J. P. Gudas, Eds, American Society for Testing and Materials, 1983, pp. II-723 - II-738.

2. W. G. Clark, Jr., 'Fracture Mechanics in Fatigue,' Experimental Mechanics, Vol. 11, Sept. 1971, p. 421.

3. N. E. Dowling, 'Fatigue Crack Growth Rate Testing at High Stress Intensities,' Flaw Growth and Fracture, ASTM ASTM STP 631, American Society for Testing and Materials, 1977, pp. 139-158.

4. V. T. Troshchenko, V. V. Pokovsky, and A. V. Prokopenko, 'Investigation of the Fracture Toughness of Constructional Steels in Cyclic Loading,' Fracture 1977, Vol. 3, 1CF7 Waterloo, Canada, June 1977.

5. J. D. Landes and J. A. Begley, 'Recent Developments in J Testing,' Developments in Fracture Mechanics Test Methods Standardization, ASTM STP 632, W. F. Brown, Jr., and J. G. Kaufman, Eds., American Society for Testing and Materials, 1977, pp. 57-81.

6. J. D. Landes and T. R. Leax, 'Load History Effects on the Fracture Toughness of a Modified 4340 Steel,' Fracture Mechanics: Fifteenth Symposium, ASTM STP 833, R. J. Sanford, Ed., American Society for Testing and Materials, Philadelphia, 1984, pp. 436-445.

7. P. K. Liaw and J. D. Landes, 'Influence of Prestrain History on Fracture Toughness Properties of Steels' Metallurgical Transactions, Vol. 17A, 1986, pp. 473-489.

8. P. K. Liaw and J. D. Landes, 'Effects of Monotonic and Cyclic Prestrain on Fracture Toughness: A Summary presented at the 18th National Symposium in Fracture Mechanics, Boulder, Colorado, June 1985 (to be published in an ASTM STP).

9. P. C. Paris and F. Erdogan, 'A Critical Analysis of Crack Propagation Laws,' Journal of Basic Engineering, Sept. 1983, p. 523.

10. Stanley T. Rolfe and John M. Barsom, Fracture and Fatigue Control in Structures, Applications of Fracture Mechanics, Prentice-Hall, Incorporated, 1977, p. 232-249.

11. N. F. Dowling and J. A. Begley, 'Fatigue Crack Growth During Gross Plasticity and the J-Integral,' Mechanics of Crack Growth, ASTM STP 590, American Society for Testing and Materials, 1976, pp. 82-103.

12. N. F. Dowling, 'Geometry Effects and th J-Integral Approach to Elastic-Plastic Fatigue Crack Growth,' Cracks and Fracture, ASTM STP 601, American Society for Testing and Materials, 1976, pp. 19-32.

13. H. A. Ernst, 'Material Resistance and Instability Beyond J-Controlled Crack Growth,' Elastic-Plastic Fracture: Second Symposium, Volume I-Inelastic Crack Analysis, ASTM STP 803, C. F. Shih and J. P. Gudas, Eds., American Society for Testing and Materials, 1983, pp. I-191-I213.

14. P. Albrecht, et al., 'Tentative Test Procedure for Determining the Plane Strain J-R Curve,' Journal of Testing and Evaluation, JTEVA, Vol. 10, No. 6, November 1982, pp. 245-251.

15. Standard Test Method for J_{IC}, A Measure of Fracture Toughness, ASTM E813-81, Annual Book of ASTM Standards, Vol. 03.01.

16. Proposed revision to the J_{IC} test standard to be released in 1987.

17. T. M. Yuzawich and C. W. Hughes, 'An Improved Technique for Removal of Oxide Scale from Fracture Surfaces of Ferrous Materials,' Practical Metallography, Vol. 15, 1978, p. 184-195.

Dislocation Cell Substructures and Fracture

R. W. Bush and D. J. Quesnel

Materials Science Program, Department of Mechanical Engineering

University of Rochester

Rochester, New York 14627

Abstract

Experimental equipment and testing procedures to investigate the effect of a cellular dislocation substructure, produced by low cycle fatigue, on the fracture behavior of a vacuum refined 0.2% carbon binary steel at -160°C are developed. Results from initial fracture tests conducted using short rod fracture specimens are given. Utilizing an appropriate analysis technique to account for plasticity, it is shown that at -160°C there is no measureable difference in fracture behavior for specimens containing a well developed dislocation substructure compared with control specimens without such structures. This is rationalized by noting that the critical cleavage event at this temperature requires little plasticity, so that differences in plastic behavior have little effect. It is anticipated that at higher temperatures the presence of a cellular dislocation substructure will have a measurable effect on the fracture behavior of this alloy.

259

Substantial effort has been devoted to the effects of prior plastic deformation on the fracture behavior of metals[1-9]. The results have been mixed, however, with some authors reporting enhanced fracture properties while others report detrimental effects or no effects at all. For instance, Groom and Knott[1] found that increasing amounts of uniform monotonic prestrain increased the local fracture stress needed to produce cleavage in mild steel, but also increased the yield stress, resulting in an increase in the transition temperature. Work by Amouzouvi and Bassim[6,7] indicates that fracture toughness (J_{IC}) of 4340 steel increases with compressive cold work up to 2% reduction in thickness, then decreases with further cold work. Liaw and Landes[9] have worked extensively on monotonic and cyclic prestrains in 4340 steel and 316 stainless steels. Their findings indicate that any prestrain history which increases material strengths will decrease fracture toughness and any prestrain history which decreases material strengths will increase fracture toughness properties.

Most of these results have been interpreted by correlating measures of material strength, such as yield stress with measures of toughness such as K_{IC}. Numerous exceptions to this broad correlation exist however. In particular, one aspect of the problem requiring further examination is the effect of the particular dislocation substructure produced by the prestrain on the fracture behavior. Since it would be expected that "anything that enhances the number of mobile dislocations, their mobility and speed, and time allowed for such movement will contribute to improved toughness"[10], then it is possible that a material with a well developed low energy dislocation structure[20], such as a cellular substructure, would exhibit superior fracture properties when compared to the same material with a higher energy dislocation structure. The work presented here is the first report of a study investigating this hypothesis.

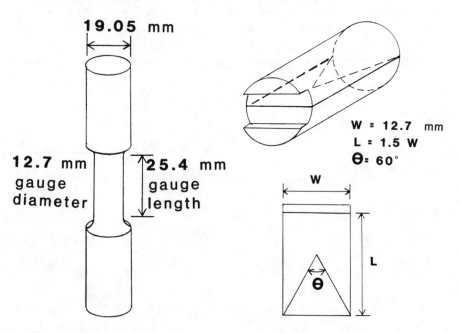

Figure 1 - Low Cycle Fatigue specimen.

Figure 2 - Short Rod specimen.

Experimental Procedure

Material

Two heats of a hot-rolled 0.20% carbon binary steel were prepared by Armco Steel Research Labs for use in this study. The chemical composition of this steel is presented in Table I. The average ferrite grain size is 30 μm, as determined by mean intercept length.[19]

Testing Procedure

Low cycle fatigue specimens were prepared as diagrammed in Figure 1. These specimens were cyclically strained with a closed loop servohydraulic testing machine, at strain amplitude of .01 and a strain rate of 5×10^{-3} sec^{-1}, at room temperature for 125 cycles where saturation was attained. The deformation was performed in a manner such that during the final loading event the material which would ultimately be deformed in tension during the fracture test was deforming in compression. Thus, the fracture test loads the material ahead of the crack tip as if it were the next cycle of cyclic

Table I. Chemical composition of Steel used in this study

Element	Sample 1	Sample 2
C	.19	.21
S	.0044	.0039
N	.0003	.0010
O	.0033	.0034
P	.006	.005
Al	.045	.043
Si	.010	.005
Mn	.053	.053
Cr	<.006	<.006
Ni	.005	.005
Ti	<.002	<.002
Sb	.005	.005
Sn	<.002	<.002
Zn	.001	.001

261

deformation. Short rod fracture specimens, illustrated in Figure 2, were then fashioned from the gauge lengths of the deformed low cycle fatigue specimens such that the crack plane was parallel with the rolling direction and perpendicular to the transverse plane. Short rod specimens were also prepared from virgin materials. All fracture specimens were 12.7mm in diameter.

Fracture specimens were tested in a low temperature chamber at a temperature of -160°C. Tests were conducted at a constant grip displacement rate of either $.0254\,^{mm}/_{sec}$ or $.0127^{mm}/_{sec}$. Load-crack opening displacement data was recorded with both an X-Y recorder and an analog to digital converter. The A-D converter was interfaced with a microcomputer, which stored the data on floppy disks for later analysis.

Crack length was measured using an unloading compliance technique. The compliance calibration, obtained previously in our laboratory[18], was performed with short rod and short bar specimens.

Experimental Equipment Description

Much of the ancillary experimental equipment used to perform these tests was designed and developed in house. This section will describe that equipment.

Figure 3 - Short Rod Specimen Saw.

Short rod specimen saw

A saw shown in Fig. 3 was necessary to cut the chevron slot into the short rod specimens. A Buehler low speed diamond saw was modified for this task. A fixture for gripping the rod at the specified chevron angle was mounted on rails enabling the rod to move vertically into the saw blade. A micrometer head mounted on the side of the rails was used to center the rod with respect to the saw blade. A counter weight filled with lead pellets ensured the proper force was applied by the specimen on the blade. It was also necessary to install a switch and rewire the leads to the motor so that the sawblade could rotate both clock wise and counter clockwise. The standard rotation direction (counter clockwise) caused binding during the slot cutting operation. Reversing the rotation alleviated this problem.

Short rod fracture test grips

The grip shown in Figure 4 is one of a pair made to conduct short rod fracture tests. They are designed so that the loading point is directly beneath the center of both the supporting rod and load cell to eliminate bending moments being tranferred to the load cell. The grip-supporting rod assembly was made in two pieces to facilitate machining, to enable us to replace a grip easily in the event of a grip failure, and to experiment with the use of different grip materials. Two grip materials were used: 304 stainless steel for low temperature work and 440C stainless steel for future higher temperature work.

Figure 4 - Specimen grip for short rod fracture testing.

Low temperature chamber

Figure 5 illustrates the low temperature chamber developed for these experiments, as well as the method of gripping the short rod specimen and attaching the extensometer to the specimen. A tank for holding liquid nitrogen was friction fit to the upper grip supporting rod to provide good thermal contact between the two pieces. A thermocouple extending approximately $\frac{1}{3}$ of the distance from the top of the tank serves as a liquid nitrogen level detector. Beneath the specimen is a nichrome heating coil which is controlled by a Variac and temperature controler in series. A type T, copper vs. copper-nickel, thermocouple was spot welded to the test specimen to provide the temperature to the controller. A second thermocouple samples the chamber air temperature as a double check. This entire assembly is surrounded by a 25 cm diameter styrofoam cylinder, which was lined with aluminum foil to prevent melting from radiant heat of the nichrome heating element. This assembly, in turn, is wrapped in two plastic encased fiber glass blankets to provide thermal insulation. A bracket was fastened to the lower grip support rod. This bracket fit into an inset on the styrofoam cylinder to prevent rotation of the chamber about the grips since the experimental set up is not axisymmetric. A 12.7 mm diameter steel specimen will reach -140° C in 1 hour and -160° C in 3 hours inside this low temperature chamber, providing a wide range of testing temperatures.

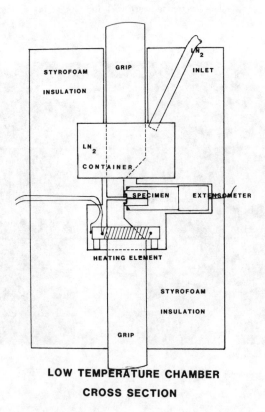

LOW TEMPERATURE CHAMBER
CROSS SECTION

Figure 5 - Diagram of low temperature chamber and specimen-extensometer attachment method.

Control Waveform

Since the crack was measured by an unloading compliance technique it was necessary to control the testing machine with a waveform that could unload partially and then reload. The testing machine's function generator could not provide this. However, by adding a ramp wave from the function generator to another lower frequency ramp wave from an auxiliary wave generator, the ncessary control waveform was produced. The two function generators had to be synchronized, however. This was accomplished as illustrated in Figure 6. The testing machine function generator is equipped with an external trigger. The output of the auxiliary wave generator was monitored with a digital voltmeter interfaced to a microcomputer via an IEEE-488 parallel port. When the output of the auxiliary wave generator crossed zero volts in a positive going manner, the microcomputer simultaneously triggered the testing machine function generator and a switch connecting the auxiliary wave generator to the "span 2" input, leading to the summing junction where the two waveforms were added to control the test. This method provides the test signal needed to allow the testing machine to partially unload and then reload at constant rate.

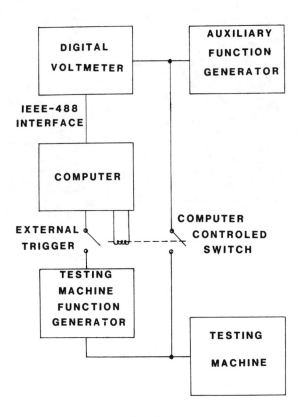

Figure 6 - Diagram of dual function generator synchronization scheme.

Data collection equipment & software.

The data was collected with an A-D converter at the rate of 20 data sets/second. A microcomputer recorded this data on floppy disk for further analysis. This analysis consists of a set of three programs written to calculate K_{IC} as a function of crack length according to the analysis described in the next section, and plot the results.

Data Analysis

Figure 7 is an example of load-crack opening displacement record which one might obtain during a fracture test. If the fracture behavior had been linear elastic, the extrapolations of all the partial unloading lines would pass through zero load at zero displacement. Since they do not, any analysis used must take into account the plasticity occuring during the fracture process.

The analysis method selected was that proposed by Burns and Michener[11] and later modified by Burns and Swain[12]. Although the theoretical approach is different, this later modification is equivalent to the method used by Barker[13] to account for plasticity in short rod specimen testing. As in linear elastic fracture mechanics, this method assumes that all irreversible energy losses have

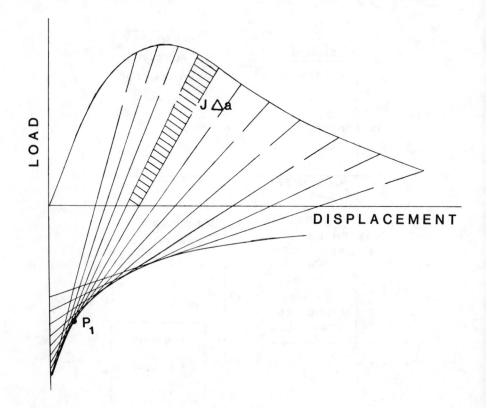

Figure 7 - Example load-crack opening displacement experimental record to illustrate the graphical data analysis technique used to determine P_1 and J.

gone into driving the crack. However, the mathematical formulation is different since a term must be added to account for the plasticity ahead of the crack tip. Simply stated, the area under the load-displacement curve bound by two adjacent unloading lines and the zero load line is equal to $J\Delta a$. Once J_{IC} is determined, the result is rewritten in terms of K_{IC} by using the equation

$$K_{IC} = \left(\frac{EJ_{IC}}{(1 - \nu^2)} \right)^{\frac{1}{2}}$$

(1)

Using the terminology of Burns and Swain

$$K_{IC} = \left(1 - 2\frac{P_1}{P} \right)^{\frac{1}{2}} P \left(\frac{E}{2B(1 - \nu^2)} \frac{dc}{da} \right)^{\frac{1}{2}}$$

(2)

where P is the applied load, E is the elastic modulus, ν is Poisson's ratio, B is the crack width, dc/da is the change of specimen compliance with crack length, and P_1 is an additional load term used to account for plasticity. This additional load term is found by using the following procedure. Figure 7 illustrates that when the partial unloading lines are extrapolated they form an envelope which is equivalent to a curve which would describe the locus of points in load-displacement space where the fracture surfaces would just come into contact. The tangent of the unloading line to this curve occurs at a load, P_1, which gives the necessary additional load term used in equation 2. The theoretical derivation of this procedure is given in Burns and Swain[12] and Michener and Burns[11].

Using this scheme, it is possible to determine K_{IC} vs. crack length as long as one ensures that all the plasticity occurs in the vicinity of the crack tip and that the plastic zone is enclosed by an elastic region, i.e. that the specimen is of sufficient size.

Resulting Data

Flow stresses and work hardening rates ($d\sigma/d\varepsilon$) were determined at a strain amplitude of .01 for low cycle fatigue geometry specimens deformed monotonically in tension and following saturation of flow stress during the final fatigue loop. Flow stress values obtained were 85 MPa and 140 MPa for virgin and cyclically deformed material respectively. Work hardening rates were 2.9GPa and 1.2GPa for virgin and cyclically deformed material respectively. These material properties serve as a measure of the differences in work hardening and yielding between the two types of specimens.

Figure 8 is a load-load point displacement trace for a fracture test conducted at -160°C. This trace is typical of what can be expected from a short rod specimen fracture test. The unloading lines

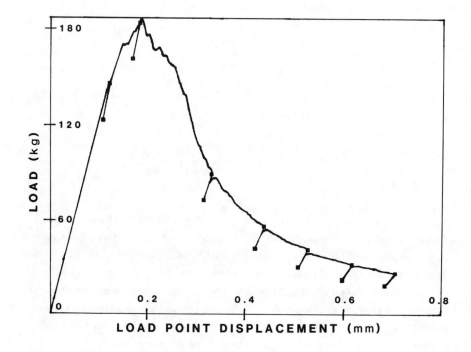

Figure 8 - Typical load-crack opening displacement experimental trace.

marked by small squares are those for which the crack length was presumed to be short enough to avoid specimen end effects.

Figure 9 shows the calculated values of K_{IC} vs. normalized crack length (crack length divided by specimen length) for cyclically deformed and virgin conditions. Two results are worthy of note. First, the cyclically deformed and virgin data do not differ significantly from one another. At first glance this is surprising since the room temperature flow stresses differ by 60%. Experimental work by Liaw & Landes[9] has shown that deformation processes which increase the flow stress of a material also decrease the fracture toughness, indicating that some difference between the cyclically deformed and virgin material is to be expected. However, the flow stresses were measured at room temperature. Data from the literature[14,15] indicates that for 1020 and mild steels the yield stress will increase about 350 MPa between room temperature and -160° C. Since this increase is independent of strain history a reasonable estimate of the yield stresses at -160° C can be obtained by adding 350 MPa to the room temperature value. This results in yield stress estimates of 435 and 490 MPa for the

268

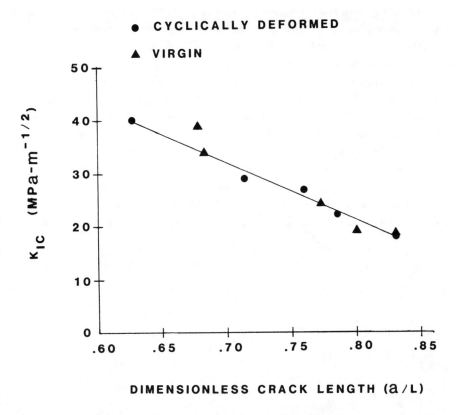

CYCLICALLY DEFORMED

VIRGIN

Figure 9 - K_{IC} vs. dimensionless crack length for cyclically deformed and virgin material at -160°C.

virgin deformed material respectively. Since this is a difference of only about 10%, if the yield stress ratio is the important parameter[17], then only a small change in fracture toughness should be expected. Such small differences would be difficult to establish in view of the measurement uncertainty.

Alternatively, it has been proposed that the critical event which triggers cleavage failure in this temperature regime is the propagation of a carbide crack into a ferrite matrix[16]. This can occur with relatively small amounts of plasticity. Therefore, at low temperatures, fracture toughness may be relatively insensitive to the dislocation structure present in the steel. Further experimentation is needed to verify this.

The second noteworthy feature of the K_{IC} vs. crack length curve is its negative slope. This is opposite to what is normally expected. As yet, we have no satisfactory explanation of this phenomenon. Further experimentation is planned to determine the origin of this observation.

Conclusions

A program to determine the effects which specific dislocation substructures have on the fracture behavior of materials is underway. Experimental equipment and protocol have been designed and developed and preliminary testing towards achieving this goal have been conducted. Also, an appropriate data analysis method has been selected and applied to account for plasticity.

Initial experiments in which Armco binary steel specimens were cycled at room temperature to produce a cellular dislocation substructure, then fractured at -160°C, and compared to fracture data of virgin specimens at -160°C have been conducted. At this low temperature there is no measurable difference between the fracture behavior of the two specimen populations. It is anticipated that differences will be found as the temperature increases since the room temperature flow stresses differ by a factor of two. At the higher temperatures, however, plasticity, not negligible even at -160°C, will play an increasingly important role in determining the fracture behavior. Using the Burns & Swain plasticity analysis, an apparent negative R curve behavior is noted in these specimens. Further experimentation is in progress to determine the origin of this observation.

References

1. J. D. G. Groom and J. F. Knott, "Cleavage Fracture in Prestrained Mild Steel," Metal Science 9 (1975), 390-400.

2. M. C. Juhas and I. M. Bernstein, "Effect of Prestrain on the Mechanical Properties of Eutectoid Steel." (Met. Trans. A., 14A (1983), 1379 - 1388.

3. Y. Mutoh, M. Toyoda, and K. Satoh, "Effect of Prestrain at Elevated Temperature on Ductile Fracture of High Strength Steel," J. Eng. Mat. Tech., 102 (1980), 364- 368.

4. K. Satoh, M. Toyoda, and Y. Mutoh, "Effect of Prestrain at Elevated Temperature on the Fracture Behavior of High Strength Steels," J. Eng. Mat. Tech., 105 (1983), 16 - 20.

5. D. A. Curry, "A Model for Predicting the Influence of Warm Prestressing and Strain Aging on the Cleavage Fracture Toughness of Ferritic Steels," Int. J. Fract., 22 (1983) 145 - 159.

6. K. F. Amouzouvi and M. N. Bassim, "Influence of Cold Working on the J-Integral in Steel," Fracture Problems and Solutions in the Energy Industry ed. L. Simpson (Oxford, England: Pergamon, 1981), 1179.

7. K. F. Amouzouvi and M. N. Bassim, "A Dislocation Model for Crack Tip Blunting in Slightly Predeformed Low Alloy Steels," Mat. Sci. Eng., 62 (1984), 137 - 146.

8. J. D. Landes and T. R. Leax, "Load History Effect on the Fracture Toughness of a Modified 4340 Steel," Fracture Mechanics, 15th Symposium ed R. J. Sanford (Philadelphia, Pa., ASTM STP 833, 1982), 436.

9. P. K. Liaw and J. D. Landes "Influence of Prestrain History on Fracture Toughness Properties of Steels, "Met. Trans. A., 17A (1986), 473 - 489.

10. R. W. Hertzberg, Deformation and Fracture Mechanics of Engineering Materials," 2nd edition, (New York, New York, John Wiley & Sons, 1983), 398.

11. J. R. Michener and S. J. Burns, "Internal Stresses in Nonlinear Fracture Mechanics." Int. J. Fracture, 23 (1983), R45 - R49.

12. S. J. Burns and M. V. Swain, "Fracture Toughness of MgO-Partially-Stabilized ZrO2 Specimens with K_R-Curve Behavior from Transformation Toughening," J. Am. Ceramic Soc., 69 (3) (1986), 226 - 230.

13. L. M. Barker, "Theory for Determining K_{IC} from Small, Non-LEFM Spcimens, Supported by Experiements on Aluminum," Int. J. Fracture, 15 (1979), 515 - 536.

14. R. O. Ritchie, J. F. Knott and J. R. Rice, "On the Relationship Between Critical Tensile Stress and Fracture Toughness in Mild Steel," J. Mech. Phys. Sol, 21 (1973), 395 - 410.

15. M. L. Wilson, R. H. Hawley and J. Duffy, "The Effect of Loading Rate and Temperature on Fracture Initiation in 1020 Hot Rolled STeel." Eng. Fract. Mech., 13 (2) (1980), 371 - 386.

16. E. Smith, The Physical Basis of Yield and Fracture, (Inst. of Physics and Phys Soc., 1966), 36.

271

17.　J. F. Knott, "Micromechanics of Fracture and the Fracture Toughness of Engineering Alloys," Fracture 1977, Volume I ed. D.M.R. Taplin (Waterloo, Ontario, Canada, University of Waterloo Press, 1977), 61 - 92.

18.　R. W. Bush and D. J. Quesnel, "Comparison of Model Material Compliance Calibration of Fracture Specimens with Results Obtained in Metals" (Paper presented at the Fall Meeting TMS-AIME, Toronto, Ontario, Canada, 14, October 1985).

19.　T. Lyman et. al. eds. Metals Handbook 8th edition, Vol. 8 (metals Park, OH: American Society for Metals, 1973), 42.

20.　M. N. Bassim et. al. eds., Low Energy Dislocation Structures, (Lausanne, Switzerland, Elsevier Sequoia, 1986).

EFFECTS OF HOT ROLLING CONDITION AND BORON-MICROALLOYING ON PHASE

TRANSFORMATION AND MICROSTRUCTURE IN NIOBIUM-BEARING

INTERSTITIAL FREE STEEL

Y. Hosoya*, S. Hashimoto** and A. Nishimoto*

* Fukuyama Research Labs., Technical Research Center, Nippon Kokan K.K.,
   Kokan-cho, Fukuyama, Hiroshima, Japan.

** Technical Research Center, Nippon Kokan K.K., Kokan-cho, Kawasaki-ku,
   Kawasaki, Japan.

Abstract
        The investigation has been made to clarify the substantial effect of
boron-microalloying on the formation of microstructure of hot band in
niobium-bearing interstitial free steel, by means of the detection of $\gamma \rightarrow \alpha$
transformation temperature, the observation of optical microstructure and
the SIMS analyses for distribution of boron-atoms. The refinement of
ferrite structure by boron-addition into niobium-bearing steel was observed
in wide range of hot rolling temperature, while the ferrite grain became
fine at the region of temperature lower than 900°C in niobium-mono-bearing
steel. The SIMS analyses revealed that the segregation of boron-atoms at
the ferrite grain boundary appeared clearly with both reduction in the
deforming temperature and increase in the amount of deformation. Effects of
these changes in the microsturcture of hot band are discussed in connection
with the mechanical properties of continuously annealed steel sheet.

Introduction

It is recognized that boron(B) is a very effective element to improve the hardenability of steel with microalloying in it by less than approximately 25 ppm(1)-(3). The most reasonable mechanism for the high hardenability in boron-bearing HSLA steel is that, boron, which is preliminarily segregated at the austenite grain boundary during soaking, reduces an interfacial energy of austenite grain boundary, and subsequently causes an inbibition of a nucleation of the ferrite at the austenite grain boundary during cooling(3). It has also been known that, on the contrary, the hardenability of boron-bearing steel was very sensitive under the heat treating conditions since the formation of some sort of precipitates, i.e., BN and $Fe_{23}(B,C)_6$ so called boron constituent, deteriorate the hardenability(2).

Due to recent progress both in the vacuum degassing process after the converter smelting and in the continuous annealing process for a production of deep drawing quality cold rolled steel sheet, in contrast to the object of boron-microalloying in HSLA steel, it has come to our attention recently that boron is also a very effective element to improve the mechanical properties of continuously annealed deep drawing quality mild steel sheet produced by an ultra-low carbon Al-killed steel(4)-(9). For example, an improvement of the strength and ductility(4)(5), an improvement of the embrittlement after press forming(4)(6), a refinement of ferrite grain of hot band(7)(8) and an improvement of the anti-aging property by precipitation of nitrogen as BN(9).

In spite of the above merits in boron-additioning in continuously annealed deep drawing quality cold rolled steel sheet, the substantial effect of boron on the formation of the hot band's microstructure in an interstitial free steel (hereinafter referred to as IF steel) has not yet been clarified, while it has been experienced that the hot rolling conditions severly affect the mechanical properties of continuously annealed IF steel sheet through a change in the microstructure of the hot band(10). To produce a deep drawing quality cold rolled steel sheet by using IF steel, accordingly, it is very important to optimize the hot rolling conditions taking a microstructural change of the hot band into consideration.

Then, in order to clarify the substantial effect of boron on the formation of the microstructure of IF steel after hot rolling by the tundem mill, the microstructural changes, the transformation temperature from austenite to ferrite and the distribution of boron after simulated hot rolling have been studied by using the ultra-low carbon steel without microalloying, the niobium-bearing ultra-low carbon steel and the niobium-bearing ultra-low carbon steel with microalloying of boron.

Experimental procedure

Three types of ultra-low carbon Al-killed steel were prepared by the converter smelting and subsequent RH degassing treatment. The chemical compositions of the steel used are shown in Table-1, i.e., No.1 : ultra-low carbon steel without microalloying, No.2 : Niobium(Nb) bearing ultra-low carbon steel and No.3 : Nb bearing ultra-low carbon steel with microalloying of boron(B).

Table 1   Chemical compositions of steel used.

(wt%)

No.	C	Si	Mn	P	S	sol.Al	N	Nb	B
1	0.0014	0.01	0.09	0.014	0.008	0.033	0.0025	-	-
2	0.0026	0.02	0.14	0.013	0.004	0.042	0.0014	0.014	-
3	0.0021	0.01	0.16	0.015	0.006	0.030	0.0017	0.019	0.0016

The starting materials for the experiment were cut from the steel bars, which were 30 mm thick, after rougher mill rolling. Cylindrical test pieces of 8 mm in diameter and 12 mm in length were prepared from the steel bars in the thickness direction, and subjected to the hot compression test according to the procedure schematically shown in Fig.1. Samples were heated at 1250°C and 1150°C for 10 min in a vacuum by means of a high frequency induction heater. After nitrogen (N_2) gas cooling in constant velocity of -50°C/sec, the first and second reductions were operated at 1020°C and 980°C, respectively. The third to fifth reduction were operated isothermally at 980°C, 950°C, 920°C, 900°C, 880°C, 870°C and 850°C with time intervals of 0.5 seconds. The strain rate of each reduction was 10 sec^{-1}. The three types of reducing schedules shown in Fig.1 were conducted in this simulation, i.e., the amount of reduction in the first compression decreases by an order from type A to type C. During N_2 gas cooling in constant velocity of -20°C/sec after the fifth reduction, the transformation temperature from the austenite (hereinafter referred to as γ) to the ferrite (hereinafter referred to as α) was determined from a dilatometric change of deformed specimen detected by means of the laser-ray delatometer.

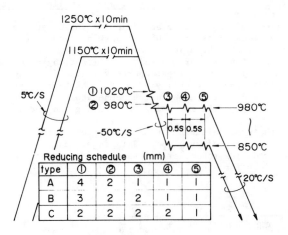

Fig.1   Schematic representation of the simulated hot compressing conditions and the reducing schedules.

Changes in the ferrite structures were observed at the central portion in a section of the specimen by an optical microscope. With respect to the B-bearing steel (steel-3), the distribution of the B-atom was observed from the 43 BO_2^- secondary ion image obtained by the irradiation of O_2^+ primary ion by SIMS (IMS-3F, CAMECA). The accelerate voltage and the analysed area were 15 kV and 150μm in diameter, respectively.

As for an estimation of the mechanical properties of continuously annealed steel sheet, three types of steels were subjected to the following test.

The block samples, dimensions of which were 100mm long, 200mm wide and 30mm thick, were cut from each steel bar and hot rolled from 30mm thick to 3.0mm thick by using the laboratory mill after soaking at 1250°C and 1150°C for 1 hour.

Final rolling temperatures were varied in the region between 950°C and 850°C taking into consideration the results obtained by the simulated hot compression test. The temperature was directly monitored by a thermo-couple preliminarily inserted into the block.

After the hot coiling simulation at 680°C for 1 hour followed by
furnace cooling, the hot bands were pickled, cold rolled to a thickness of
0.8mm and annealed at 820°C for 90 sec as a simulation of continuous
annealing. Finally, tensile test pieces, dimensions of which were 25mm in
gauge width and 50mm in gauge length, were machined from the annealed sheets
after temper rolling with a reduction of 0.6%. The Lankford-values
(r-values) were measured for longitudinal, diagonal and transverse
directions at 20% tensile strain. Mean r-value was assessed by the
following equasion.

Mean r value = $(r_L + 2r_D + r_T)/4$
r_L : r value in longitudinal direction
r_D : r value in diagonal direction
r_T : r value in transverse derection
Aging index was measured by an increment of the flow stress by
artificial aging at 100°C for 1 hour after tensile straining up to 8%.

Results

Change in the ferrite structure caused by the hot rolling conditions in
three types of steel.
    The optical microstructures observed in the three types of steel, which
were tested by the conditions of 1150°C in soaking temperature and type A
in reducing schedule, are shown in Photo.1 as a typical case. As compared
with the grain size of ultra-low C steel, the microalloying of Nb causes a
marked refinement of α grain at the temperature which is lower than 900°C.
By means of the additional microalloying of B in Nb-bearing steel, however,
the α grain is refined stably even in the temperature which is higher than
900°C. And in this case, the morphology of the α grain boundary is apt to
become wavy.
    Changes in the grain size number as a function of the final reducing
temperature (hereinafter referred to as FRT) in all cases of soaking
temperatures and reducing schedules are shown in Fig.2 for steel-1, Fig.3
for steel-2 and Fig.4 for steel-3.

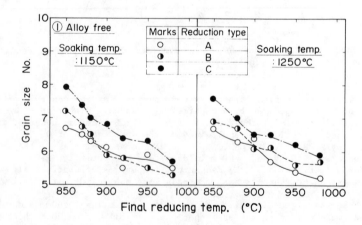

Fig.2   Changes in the grain size number of ferrite as a function of the
         final reducing temperature in steel-1

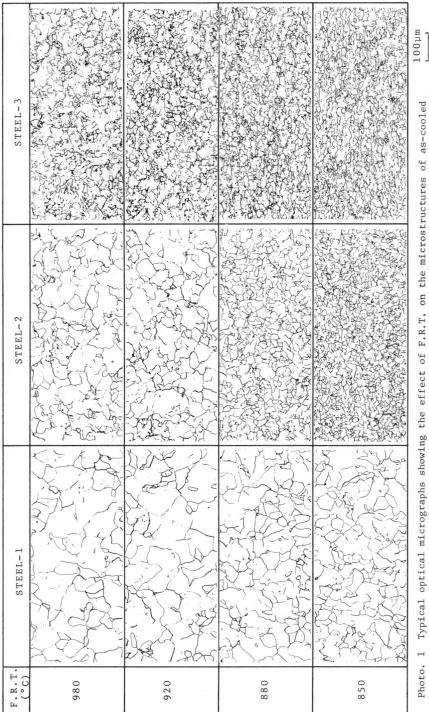

F.R.T. (°C)	STEEL-1	STEEL-2	STEEL-3
980			
920			
880			
850			

100μm

Photo. 1 Typical optical micrographs showing the effect of F.R.T. on the microstructures of as-cooled samples observed in STEEL-1, STEEL-2 and STEEL-3. (Soaking : 1150°C, Reduction schedule : A)

277

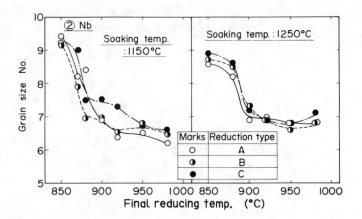

Fig.3　Changes in the grain size number of ferrite as a function of the
final reducing temperature in steel-2.

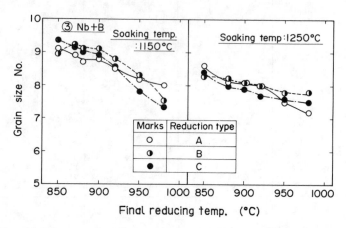

Fig.4　Changes in the grain size number of ferrite as a function of the
final reducing temperature in steel-3.

　　　As for steel-1, the grain size number of α gradually becomes fine, not
only by a lowering of FRT, but also by an increase in the amount of
reduction at FRT.　In this case, refinement of the α grain is considered to
be directly affected by a refinement of the γ grain.
　　　As for steel-2, however, the α grain is markedly refined with reducing
FRT to the region of temperature which is lower than 900°C.　Even in the
ultra-low C steel, it is considered that Nb acts as a potent inhibitor of
the recrystallization of γ at low rolling temperature, so this causes the
increase in the γ grain boundary area and the introduction of deformation
band which enhance the nucleation of α grain(11).
　　　As for steel-3, an additional microalloying of B in Nb-bearing
ultra-low C steel refines the α grain, which is finer than 7.5 in grain size
number, even in the case of high FRT such as 980°C.　It is the
characteristics of the α grain refinement in steel-3 that the grain size of α
is mainly affected not by the reducing schedule but by the lowering of FRT.

In order to compare the absolute levels of grain size among the three
types of steel, and to confirm the effect of soaking temperature on the
grain size in each steel, the results obtained with respect to the reducing
schedule A are shown in Fig.5.

In steel-1, a significant difference in the α grain size, which is
caused by a change in soaking temperature, is observed when FRT is higher
than 900°C. The vanishing of the effect of the change in soaking
temperature on the α grain size, which is observed at low rolling
temperature, suggests a convergence of the grain size of recrystallized γ
due to severe deformation at low FRT despite that the prior γ grain sizes
are different between two conditions in soaking.

In contrast to steel-1, it is observed in steel-2 that a high soaking
temperature is effective for a refinement of the α grain. This is
considered to be caused by the fact that the dragging of solute Nb and the
fine precipitation of NbC during hot deformation causes the refinement of the
dynamically recrystallized γ grain, since NbC is sufficiently dissolved by
the high temperature soaking. With respect to a deep drawing quality cold
rolled steel sheet, however, the fine distribution of carbonitride composed
of the microalloying element, such as NbC in this experiment, badly affects
the recrystallization texture after annealing(10). So, the stable grain
refinement of α is expected by low soaking temperature. Additional
microalloying of B enables to make the α grain be fine by soaking at low
temperature.

This reversion in the effect of soaking temperature on the α grain size
between steel-2 and 3 seems to be closely related to the segregation of B at
the γ grain boundary. That is to say, the finer a primary γ grain is by the
soaking at low temperature, the finer the distribution of B becomes. This
phenomenon is to be discussed later taking the results on the $\gamma \rightarrow \alpha$
transformation into consideration.

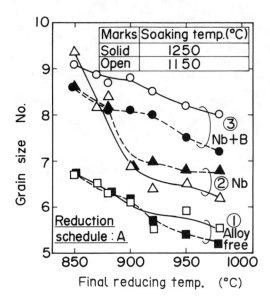

Fig.5   Comparison among three types of steels on the changes in the grain
        size number as a function of the final reducing temperature due to
        difference in the soaking temperature.

$\gamma\rightarrow\alpha$ phase transformation after hot rolling in three types of steel.

The starting and finishing temperatures of the $\gamma\rightarrow\alpha$ transformation after hot rolling were detected by a delatometric change of specimen during cooling in constant velocity (hereinafter, the starting and finishing temperatures are referred to as Ts and Tf, respectively).

Dilatometric changes during cooling, observed in the case in which the soaking temperture was 1150°C and the reducing schedule was type A, are shown in Fig.6 for steel-1, Fig.7 for steel-2 and Fig.8 for steel-3.

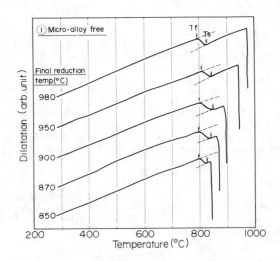

Fig.6　Typical dilatometric changes during cooling observed in steel-1 tested by the combination of the soaking at 1150°C and the reducing by schedule A.

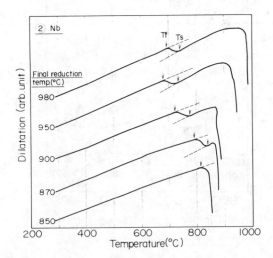

Fig.7　Typical dilatometric changes during cooling observed in steel-2 tested by the combination of the soaking at 1150°C and the reducing by schedule A.

280

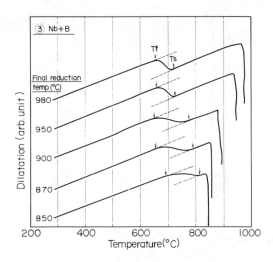

Fig.8　　Typical dilatometric changes during cooling observed in steel-3
tested by the combination of the soaking at 1150°C and the reducing
by schedule A.

The degree of supercooling of Ts from Ar_3 transformation temperature
in each type of steel is in the same order to the results reported by
Serin(12) in the same alloying system, i.e., C-Mn, C-Mn-Nb, and C-Mn-Nb-B.
However, the absolute levels of Ts are higher than the results obtained by
Serin because C and Mn contents are very low and the transformations are
enhanced by the hot defomation in this experiment.
　　The point of special mention in the dilatometric change due to $\gamma \rightarrow \alpha$
transformation in three types of steel is that the rate of expansion during
the $\gamma \rightarrow \alpha$ transformation is markedly decelerated by lowering the FRT only in
the case of steel-3.
　　Then the changes in both Ts and Tf as a function of FRT in the three
types of steel are examined taking into consideration the differences in
both soaking temperature and reducing schedule.　In order to clarify the
change in the temperture of the $\gamma \rightarrow \alpha$ transformation due to final reduction,
Figs. 9 and 10 show the change in Ts and Tf in the $\gamma \rightarrow \alpha$ transformation
obtained during cooling from the conditions of as-soaked and of first and
second reductions.
　　As compared to the static transformation detected during cooling from
the as-soaked condition in each steel, the change in the temperature of the
$\gamma \rightarrow \alpha$ transformation after first and second reductions are as follows.
　　In steel-1, Ts and Tf after first and second reductions hardly change
compared to those in the static transformation.
　　In steel-2, marked reduction in Tf is observed in comparision with that
in steel-1.　However, first and second reductions cause an ascension of Tf,
and result in the narrowing of the region between Ts and Tf in the $\gamma \rightarrow \alpha$
transformation.　Judging from the result that this tendency is more obvious
in the reducing type A, it is considered that the hot deformation enhances
the $\gamma \rightarrow \alpha$ transformation even under the conditions that the total reduction is
less than 50% and the final reducing temperture is higher than 980°C.

281

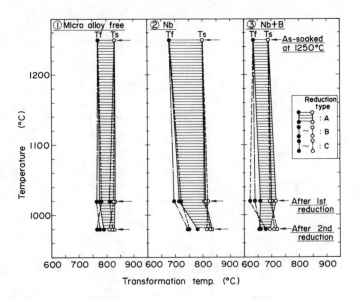

Fig.9   Changes in both Ts and Tf in the $\gamma \rightarrow \alpha$ transformation during cooling
from the soaking temperature of 1250°C and from the subsequent
first and second reductions.

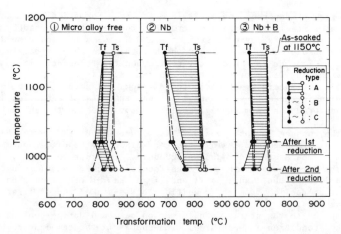

Fig.10   Changes in both Ts and Tf in the $\gamma \rightarrow \alpha$ transformation during cooling
from the soaking temperature of 1150°C and from the subsequent
first and second reductios.

In steel-3, both Ts and Tf are markedly reduced in the static
transformation. This tendency is hardly affected by the second reduction.
This result is considered to be caused by an inhibition of the α nucleation
at the γ grain boundary due to a segregaion of B(3).

Subsequently, the changes in Ts and Tf after final reduction examined
by soaking at 1250°C and 1150°C are shown in Fig.11 and Fig.12,
respectively.

282

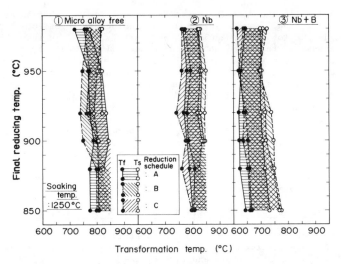

Fig.11   Changes in both Ts and Tf in the $\gamma \to \alpha$ transformation during cooling
after the final reduction at various temperature.
(Soaking temperature: 1250°C)

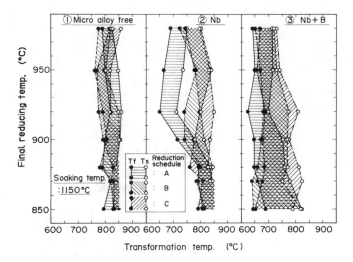

Fig.12   Changes in both Ts and Tf in the $\gamma \to \alpha$ transformation during cooling
after the final reduction at various temperature.
(Soaking temperature: 1150°C)

With respect to steel-1, the region of the temperature in the $\gamma \to \alpha$
transformation is hardly dependent on both FRT and the reducing schedule.
By microalloying of Nb into an ultra-low carbon steel, that is steel-2,
Tf ascends still more in comparison to that observed after second reduction.
This tendency is more obvious in the case of soaking at 1250°C, the case of
which is that the NbC particles completely dissolve during soaking
calculated by the solubility limit of NbC in the γ phase(13).

283

In the case of soaking at 1150°C, the condition of which is considered to be incomplete in the dissolution of NbC by soaking for 10 minutes, the reductions in Ts and Tf are observed at the region of temperature higher than 920°C in FRT for the schedule A. However, Ts and Tf in the schedule A markedly ascends to the levels obtained in schedules B and C by the final reduction at the temperature which is lower than 900°C.

Marked enhancement of the $\gamma \to \alpha$ transformtion after the hot deformation, observed in Nb-bearing IF steel, is considered to be caused by the strain induced α nucleation at the deformed γ grain boundary and the deformation band, and this tendency is closely related to the refinement of the α grain as shown in Fig.3.

In steel-3, the expansion of the region between Ts and Tf, as compared to the static transformation, is observed after the final reduction. Especially, the ascension of Ts is remarkable while Tf changes minimally. This change is more obvious in the case of soaking at 1150°C than that of soaking at 1250°C. Additionally, both Ts and Tf ascend with an increase in the amount of deformation in the final reduction.

Observation of boron distribution in B-bearing IF steel after hot deformation.

With respect to the detection of the distribution of B in steel, secondary ion mass spectrometry (SIMS) has been applied as an effective technique(14)(15)(16). In this experiment, the distribution of B in steel-3 was observed as BO_2^- secondary ion image by SIMS(16).

Photo.2 shows the secondary ion images observed in the samples which were finally reduced at 980°C, 920°C, 880°C and 850°C in the reducing schedules A and C after 1150°C soaking.

Numerous spot images, which are considered to be the precipitates composed of BN, are observed in every samples. In addition, the network images come to be observed in the conditions of low FRT in both types of reducing schedules. Especially in both cases of 850°C in FRT, network images are very obvious. This image is considered to be from B which segregated at the grain boundary.

Photo.3 shows the optical micrograph and the BO_2^- secondary ion image by same magnification observed in the sample, finally compressed at 850°C by schedule C.

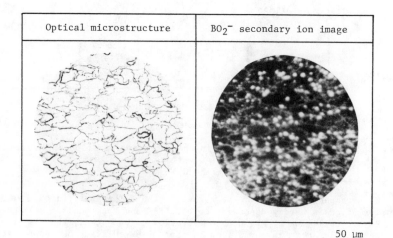

| Optical microstructure | BO_2^- secondary ion image |

50 µm

Photo. 3   Comparison between optical microstructure and BO_2^- secondary ion image in the sample finally compressed at 850°C by reducing schedule C after soaking at 1150°C.

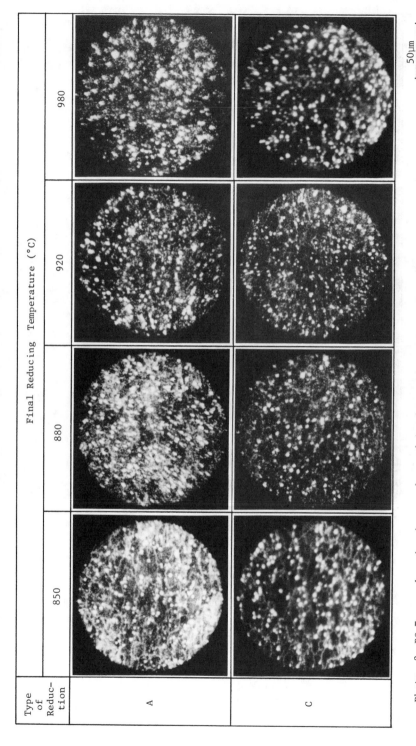

Photo. 2 $BO_2{}^-$ secondary ion images observed in the samples which were finally reduced at 980°C, 920°C, 880°C and 850°C in reduction schedules A and C after soaking at 1150°C.

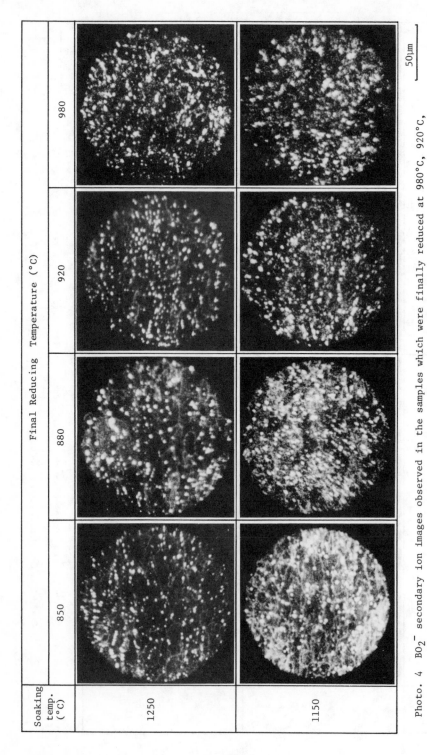

Photo. 4 BO$_2^-$ secondary ion images observed in the samples which were finally reduced at 980°C, 920°C, 880°C and 850°C in reduction schedule A after soaking at 1250°C and 1150°C.

The size and the morphology of the network structure in the secondary ion image are very similar to those of the optical microstructure. By a rough estimation of the grain size number of network images, about 9.0 and 9.6 are obtained in the samples finally compressed at 850°C by schedules A and C, respectively. These values are very close to the grain size number of the α as shown in Fig.4. The problem of whether this network image corresponds to either the γ grain boundary or the α grain boundary is to be addressed in Discussion.

The effect of soaking temperature on the distribution of B observed in the samples deformed by the reducing schedule A are shown in Photo.4.

In the case of the soaking at 1250°C, the network images are observed in the samples finally reduced at FRT lower than 920°C. Grain size number which was roughly estimated from the image in the sample which was soaked at 1250°C and finally reduced at 850°C is about 8.5. This value is also very close to the grain size number of the α as shown in Fig.4.

Effect of final reducing temperature (FRT) on the mechanical properties of continuously annealed steel sheet

In order to demonstrate the effect of microstructural change in hot band due to B-microalloying on the mechanical properties of continuously annealed steel sheet, the changes in yield strength, tensile strength, work hardening exponent, mean r-value and aging-index of continuously annealed three types of steel sheet are shown in Fig.13 as a function of the FRT monitored.

From the results on the aging-index, additional microalloying of B into the Nb-bearing ultra-low carbon steel is very effective for the reduction of a strain aging by an entire precipitation of nitrogen as BN.

On the mean r-value, it in steel-1 decreases with the reduction in FRT. This is caused by the hot rolling at the inter-critical temperature below Ar$_3$. As for steels-2 and 3, relatively high r-values are obtainable by a reduction in the carbon in solution. However, the decrease in the mean r-value at FRT lower than 900°C, observed in steel-3 soaked at 1250°C, indicates that the mean r-value is deteriorated by the hot rolling at low FRT which causes the dynamic B-segregation at the α-grain boundary as shown in Photo.4. This is considered to be related to the change in the cold rolling texture due to segregation of B at the α-grain boundary in hot band.

Consequently, from a view point of the refinement of the hot band's microstructure without occurrence of the B-segregation at the α grain boundary, 1150°C soaking is better than 1250°C soaking. And in this case, microalloying of B up to 16 ppm into Nb-bearing ultra-low carbon steel hardly deteriorate the mean r-value of continuously annealed sheet.

With respect to the yield strength, the tensile strength and the work hardening exponent, low FRT causes an increase in the strength levels and a decrease in the uniform elongation which is assessed by the work hardening exponent. This is considered to be caused by the lack of grain growth during recrystallization in a short time annealing due to existance of the B-atom preliminarily segregated at the α grain boundary in hot band.

Accordingly, the deep drawing quality cold rolled steel sheet characterized with non-aging, high r-value, good ductility and having enough yield and tensile strength levels is produceable in the continuous annealing process by using the Nb-bearing ultra-low carbon steel with microalloying of B, if the hot rolling conditions are optimized as follows.

1. Soaking temperature is low in order to control the dissolution of NbC and BN in hot band.

2. FRT is controlled to approximately 900°C in order to avoid the B-segregation at the α-grain boundary caused by the increase in the strain energy during hot rolling due to both deformation in low FRT and big reduction at the latter half of the rolling schedule.

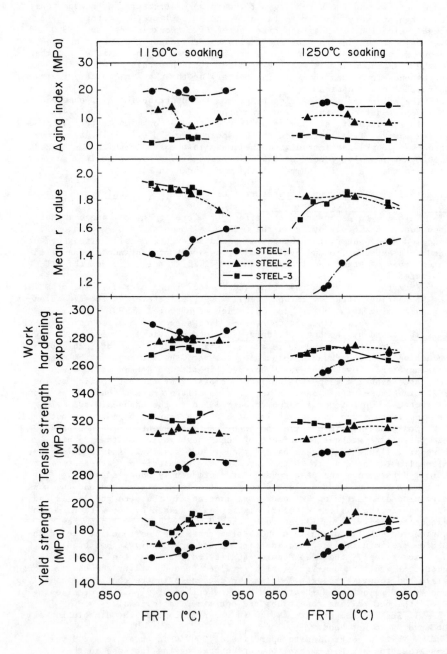

fig.13   Effects of the soaking temperature and FRT in the hot rolling on
the mechanical properties of the steel sheets continuously annealed
after cold rolling.

Discussion

It has become clear that the microalloying of B into Nb-bearing ultral-low carbon Al-killed steel led to the marked refinement of the α grain, and this tendency was considered to be closely related to the process of the $\gamma \to \alpha$ transformation after hot deformation.

Watanabe(17) had studied the influence of hot rolling on the distribution of B in steel with a view to adopting the B-bearing HSLA steel to the direct quenching process, and proposed the process of the dynamic segregation of B at the γ grain boundary after hot rolling.

Figure 14 shows the schematic representation of the dynamic segregation of B at the γ grain boundary. The process of which is summarized as follows.

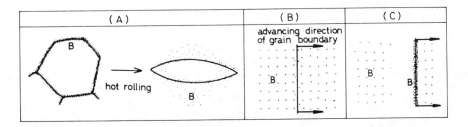

Fig.14   Schematic representation of the distribution of boron atoms, (A): during hot rolling where moving velocity of boundary is much higher than boron atoms, (B): just after hot deformation and (C): after boundary migration due to recrystallization(14).

As a primary condition, B segregates at the γ grain boundary during soaking. Under the hot rolling in the strain rate from 10 to 70 sec^{-1}, B is left behind the moving γ grain boundary because the diffusion rate of B is slower than the moving velocity of the γ grain boundary (process-(A)). After hot rolling, however, B is able to follow the advancing recrystallyzed grain boundary, and segregates at the γ grain boundary (processes-(B) and (C)).

Then, the results obtained in this investigation with respect to the behavior of B after hot deformation are discussed below in comparison with the dynamic segregation mechanism of B as shown in Fig.14.

The segregation of B at the primary γ grain boundary during soaking is also suggested from the results on the inhibition of the $\gamma \to \alpha$ transformation in the static condition as shown in Figs. 9 and 10. And the strain rate of 10 sec^{-1}, which was the subjected condition in this experiment, is probably enough to leave the B atom behind the moving γ grain boundary under hot deformation.

The significant point in this study is that, while the high hardnable HSLA steel was water quenched from the γ phase followed by martensite transformation in Watanabe's work, the ultra-low carbon IF steel with scarce hardenability was used and the $\gamma \to \alpha$ transformation occured during cooling in the velocity of $-20°C/sec$.

Therefore, it must be considered that the α grain nucleates at both the recrystallized γ grain boundary and the deformation band with diffusion of interstitial atoms.

Figure 15 shows the schematic representation on the process of the $\gamma \to \alpha$ transformation after hot deformation proposed by the present authors.

Since the solubility of B in the α-phase is much less than the γ-phase(18), B which is in both solute condition and dynamically segregated at the γ grain boundary is considered to be excluded at the front of the advancing γ/α interface during the $\gamma \to \alpha$ transformation, and the dragging of the cluster of segregated B probably causes the resistance for the movement of ℓ/α interface. Finally, B-atom segregates at the α grain boundary just after the $\ell \to \alpha$ transformation. This process is suggested from the following results. 1. SIMS observation has revealed that the B distributes as a network structure, and the size of which was very similar to the α grain size. 2. A marked deceleration of the $\ell \to \alpha$ transformation, i.e., the expansion of the region of temperature between Ts and Tf, was observed in the samples which contained the B segregated at the grain boundary. 3. The nucleation of the α grain was not inhibited but promoted in the samples which contained the B segregated at the grain boundary in contrast to the static $\ell \to \alpha$ transformation without hot deformation.

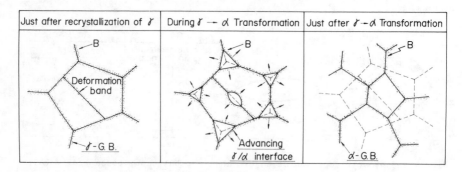

Fig.15   Schematic representation of the distribution of boron atoms during the $\ell \to \alpha$ transformation after hot deformation speculated from the results obtained by the B-bearing ultra-low carbon steel (steel-3).

The effects of the reducing schedule and soaking temperature on the B-segregation, which were shown in Photos. 2 and 4, respectively, are considered to be caused by the following mechanisms in view of the process of the B-segregation at the α grain boundary mentioned above.

In the condition of severe reduction in the latter half of the reducing schedule (type C), the α-nucleation inside the ℓ grain is accelerated not only by an activation of nucleation but also by an increase in the number of nucleation sites. Accordingly, it is speculated that B rapidly segregates at the ℓ/α interface by both decrease in the mean free path of the nucleus and a promotion of diffusion of B by a short cut diffusion through a lattice defects, and behaves as a dragging force for the movement of ℓ/α interface.

On the effect of soaking temperature, since the ultra-low carbon steel was used in this experiment, the amount of B precipitated as boron constituent(2) was probably less than the amount of the increased B in solution from dissolved BN by soaking at high temperature. Therefore, it is considered that the increase in B as an interstitial atom promotes the segregation of B at the α grain boundary at higher a temperature in FRT. And it is considered that this difference in the amount of B in solution due to change in the soaking temperature causes the deterioration of the r-value and the work hardening exponent of continuously annealed steel sheet as shown in Fig.13.

According to the above discussion, the effects of microalloying elements and the severity of hot deformation on the $\ell \to \alpha$ transformation temperature are schematically represented in Fig.16.

In the case of the ultra-low carbon steel without microalloying, the $\ell \to \alpha$ transformation temperature is hardly affected by the changes in FRT and the reducing schedule conducted in this experiment.

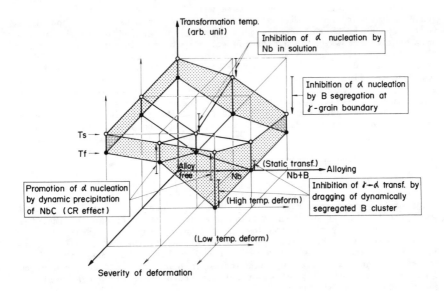

Fig.16   Schematic diagram showing the effects of both microalloying of Nb
and B into an ultra-low carbon steel and severity of hot
deformation on Ts and Tf in the $\gamma \to \alpha$ transformation.

In the case of the Nb-bearing ultra low carbon steel, the α-nucleation
is markedly enhanced with an increase in the severity of deformation caused
by the dynamic precipitaion of fine NbC and the introduction of the
deformation band, while the α-nucleation is inhibited by the existance of Nb
in solution in the static transformation. In the case of the Nb-bearing
ultra low carbon steel with microalloying of B, the $\gamma \to \alpha$ transformation in
static conditon is marked inhibited by the segregaion of B at the primary
γ-grain boundary. With respect to the $\gamma \to \alpha$ transformation after hot
deformaion, however, the rate of the $\gamma \to \alpha$ transformation is decerelated by
the dragging of dynamically segregated B at the γ/α interface.
However, the degree of B-segregation at the α-grain boundary in hot
band should be controlled by the optimization of the hot rolling conditions
taking into consideration the deterioration of mechanical properties of
continuously annealed steel sheet due to B-segregation.

Conclusion

Microstructural changes in the hot band and the mechanical properties
of continuously annealed steel sheet of interstitial free steel with
microalloying of niobium, and with complex microalloying of niobium and
boron were studied from the results of the changes in the $\gamma \to \alpha$ phase
transformation temperature and the distribution of boron in steel.
Conclusions are summarized as follows.
1   In the ultra-low carbon steel without microalloying, the temperature of
the $\gamma \to \alpha$ phase transformation is hardly affected by the change in the final
reducing temperature (FRT) within the region from 980°C to 850°C.
However, the α-grain is refined to some degree corresponding to the
refinement of the γ-grain with lowering FRT.

291

2   The α-grain of Nb-bearig ultra-low carbon steel is markedly refined by the hot deformation at the temperature which is lower than 900°C. The grain refinement by the microalloying of Nb into an ultra-low carbon steel is mainly caused by the accerelation of the α-nucleation inside γ grains not only by the activation of nucleation but also by the increase in the number of nucleation site due to severe hot deformation. And this tendency is closely related to the rise in the $\gamma \rightarrow \alpha$ transformation temperature with lowering the final reducing temperature.

3   By the additional microalloying of B into Nb-bearing ultra-low carbon steel, the α-grain is markedly refined in the wide range of the final reducing temperature from 980°C to 850°C, and the finishing temperature of the $\gamma \rightarrow \alpha$ transformation (Tf) hardly increases with both lowering the final reducing temperature (FRT) and increasing the amount of reduction at the latter half of reducing schedule, despite that the starting temperature of the transformation markedly ascends.

4   SIMS observation revealed the segregation of B at the α-grain boundary after hot deformation. This is speculated to be the consequence of a strain induced dynamic segregation of B atom at the γ/α interface in the early stage of the phase transformation followed by a dragging of the segregated B-cluster to the end of the $\gamma \rightarrow \alpha$ transformation. It is considered that the grain refinement by the microalloying of B into Nb-bearing ultra-low carbon steel is caused by the dragging of B-cluster during the $\gamma \rightarrow \alpha$ transformation.

5   The B-segregation at the α-grain boundary in hot band due to deformation at low FRT which was observed in B-bearing IF steel, causes the deterioration of the mechanical properties of continuously annealed steel sheet. Consequently, the hot rolling conditions for the B-bearing IF steel should be the combination of a low soaking temperature, such as 1150°C, and the optimized final reducing temperature, which is low enough to refine the α-grain without the occurence of the B-segregation at the α-grain boundary, i.e, approximately 900°C.

References

(1) R. A. Grange and T. M. Garvey, "Factors Affecting the Hardenability of Boron-Treated Steels", Trans. ASM, vol.37 (1946), PP.136-174.

(2) G. F. Melloy, P. R. Slimmon and P. P. Podgursky, "Optimizing the Boron Effect", Met. Trans., vol.4 (1973), PP.2279-2289.

(3) Ph. Maitrepierre, J. Rofes-Verris and D.Thivellier, "Structure-Properties Relationships in Boron Steels", PP.1-18 in "Boron in Steel" ed. by S. K. BANERJI and J. E. MORRAL, TMS-AIME, Milwaukee, Wisconsin, 1979.

(4) N. Takahashi, M. Shibata, Y. Furuno, H. Hayakawa, K. Kakuta and K. Yamanoto, "Boron-Bearing Steels for continuous Annealing to Produce Deep-Drawing and High-Strength Steel Sheet", PP.133-153 in "Metallurgy of Continuous-Annealed Sheet Steel", ed. by B. L. BRAMFITT and P. L. MANGONON, JR., TMS-AIME, Dallas, Texas, 1982.

(5) R. Pradhan, "Effect of Nitride Formers (B, Z, Ti) on the Mechanical properties of Continuously Anneled Low-Carbon Steel Sheet", PP.185-202 in "Technology of continuously annealed Cold Rolled Sheet Steel", ed. by R. Pradhan, TMS-AIME, Detroit, Michigan, 1986.

(6) T. Suda, M. Sakoh, K. Tayama, K. Araki, M. Wada and H. Kajitani, "Improvement of Embrittlement after Press Forming in Nb-bearing Ultra-Low Carbon Steel", Tetsu-to-Hagané, vol.69 (1983), S1127.

(7) Y. Ito, M. Nakazawa, Y. Nakazato and N. Ohashi,"Deep-drawing Quality Hot Rolled Steel Sheet KFN", Kawasaki Steel Technical Report, vol.5 (1973), PP.224-234.

(8) H. Naemura, L. Nozoe, M. Jitsukawa, T. Shimomura and S. Ono, "Production of Deep-Drawing Quality Cold Rolled Steel Sheet", Tetsu-to-Hagané, vol.67 (1981), S1178.

(9) M. Sudo and I. Tsukatani, "Effect of Aluminum Addition on the Mechanical Properties of Boron-Bearing Galvanized Sheet Steels", PP.203-218 in "Technology of Continuously Annealed Cold-Rolled Sheet Steel", ed. by R. Pradhan, TMS-AIME, Detroit, Michigan, 1986.

(10) S. Satoh, T. Obara, M. Nishida and T. Irie, "Effects of Alloying Elements and Hot-Rolling Conditions on the Mechanical Properties of Continuous-Annealed, Extra-Low-Carbon Steel Sheet", PP.151-166 in "Technology of Continuously Annealed Cold-Rolled Sheet Steel", ed. by R. Pradhan, TMS-AIME, Detroit, Michigan, 1986.

(11) R. Priestner, "Strain-Indnced $\delta - \alpha$ Transformation in the Roll Gap in Carbon and Microalloyed Steel", PP.455-466 in "Thermomech Process Microalloyed Austenite", (1982).

(12) B. Serin, Y. Desalos, Ph. Maitrepierre and J. Rofes-Vernis, "Caractéristiques de transformation et propriétés daciers à bas carbone au Nb-B", Menoires Scientifiques Revue Metallurgie Juin (1978), PP.355-369.

(13) T. H. Johansen, N. Christensen and B. Augland, "The Solubility of Niobium Carbide in Gamma Iron", Trans. Met. Soc. AIME, vol.239 (1967), PP.1651-1654.

(14) T. B. Cameron and J. E. Morral, "Methods of Detecting Boron in Steel", PP.61-79 in "Boron in Steel" ed. by S. K. BANERJI and J. E. MORRAL, TMS-AIME, Milwaukee, Wisconsin, 1979.

(15) T. Shiraiwa, N. Fujino and J. Murayama, "Ion Microprobe Mass Analysis of boron in Steel", Tetsu-to-Hogané, vol.67 (1981), PP.606-182.

(16) S. Hashimoto, S. Doi, M. Terasaka, K. Takahashi and M. Iwaki, "Quantitative Analysis of Boron in Steel by SIMS using Ion-Implanted Iron as Standards", Materials Science and Engineering, (1987), to be published.

(17) S. Watanabe, H. Ohtani and T. Kunitake, "The Influence of Hot Rolling and Heat Treatments on the Distribution of Boron in Steel", Tetsu-to-Hagané, vol.62 (1976), PP.1842-1850.

(18) Y. Hayashi and T. Sugeno, "Nature of Boron in α-Iron", Acta Met., vol.18 (1970), PP.693-697.

Subject Index

Author Index